Handbook of
Cardiotocography

Handbook of
Cardiotocography

Editor

Roza Olyai MS (Obstetrics & Gynaecology) FICOG FICMCH

Vice President, Family Planning Association of India (FPAI) 2018–20
Member, South Asian Regional Governing Council—International Planned Parenthood (IPPF) 2017–20
Member, Governing Council Indian College of Obstetrician & Gynaecologist (ICOG) 2018–20
Vice President, Federation of Obstetric & Gynaecological Societies of India (FOGSI) 2014
Chairperson, Adolescent Health Committee (FOGSI) 2009–12
Director, Olyai Hospital, Hospital Road, Gwalior, MP, India

Foreword

CN Purandare

CBS

CBS Publishers & Distributors Pvt Ltd

New Delhi • Bengaluru • Chennai • Kochi • Kolkata • Mumbai
Hyderabad • Jharkhand • Nagpur • Patna • Pune • Uttarakhand

Disclaimer

Science and technology are constantly changing fields. New research and experience broaden the scope of information and knowledge. The editor has tried her best in giving information available to her while preparing the material for this book. Although, all efforts have been made to ensure optimum accuracy of the material, yet it is quite possible some errors might have been left uncorrected. The publisher, the printer and the editor will not be held responsible for any inadvertent errors, or inaccuracies.

Handbook of
Cardiotocography

ISBN: 978-93-87085-90-9

Copyright © Publisher

First Edition: 2018

Published by Satish Kumar Jain and produced by Varun Jain for

CBS Publishers & Distributors Pvt Ltd
4819/XI Prahlad Street, 24 Ansari Road, Daryaganj, New Delhi 110 002, India.
Ph: 23289259, 23266861, 23266867   Website: www.cbspd.com
Fax: 011-23243014   e-mail: delhi@cbspd.com; cbspubs@airtelmail.in.

Corporate Office: 204 FIE, Industrial Area, Patparganj, Delhi 110 092
Ph: 4934 4934   Fax: 4934 4935   e-mail: publishing@cbspd.com; publicity@cbspd.com

Branches

- **Bengaluru:** Seema House 2975, 17th Cross, K.R. Road, Banasankari 2nd Stage, Bengaluru 560 070, Karnataka
  Ph: +91-80-26771678/79   Fax: +91-80-26771680   e-mail: bangalore@cbspd.com
- **Chennai:** 7, Subbaraya Street, Shenoy Nagar, Chennai 600 030, Tamil Nadu
  Ph: +91-44-26680620/26681266   Fax: +91-44-42032115   e-mail: chennai@cbspd.com
- **Kochi:** Ashana House, No. 39/1904, AM Thomas Road, Valanjambalam, Ernakulam 682 016, Kochi, Kerala
  Ph: +91-484-4059061-65   Fax: +91-484-4059065   e-mail: kochi@cbspd.com
- **Kolkata:** 6/B, Ground Floor, Rameswar Shaw Road, Kolkata-700 014, West Bengal
  Ph: +91-33-22891126, 22891127, 22891128   e-mail: kolkata@cbspd.com
- **Mumbai:** 83-C, Dr E Moses Road, Worli, Mumbai-400018, Maharashtra
  Ph: +91-22-24902340/41   Fax: +91-22-24902342   e-mail: mumbai@cbspd.com

Representatives

• **Hyderabad**	0-9885175004	• **Jharkhand**	0-9811541605	• **Nagpur**	0-9021734563
• **Patna**	0-9334159340	• **Pune**	0-9623451994	• **Uttarakhand**	0-9716462459

Printed at: HT Media, Noida, UP, India

to

My mother in-law

Late Professor Dr (Mrs) Perin Olyai

Professor Emeritus

Ex Head of the Department of Obstetrics and Gynaecology

Gajra Raja Medical College, Gwalior, Madhya Pradesh, India

For her genuine love and dedication towards service to humanity

Foreword

The evidence for the benefits of continuous CTG monitoring, as compared to intermittent auscultation, in both low and high-risk labours is scientifically inconclusive. The use of continuous intrapartum CTG in low-risk women is more controversial, although it has become standard of care in many countries.

When compared to intermittent auscultation, continuous CTG has been shown to decrease the occurrence of neonatal seizures, but no effect has been demonstrated on the incidence of overall perinatal mortality or cerebral palsy.

Most experts believe that continuous CTG monitoring should be considered in all situations where there is a high-risk of fetal hypoxia/acidosis, whether due to maternal health conditions, abnormal fetal growth during pregnancy, epidural analgesia, meconium stained liquor, or the possibility of excessive uterine activity, as occurs with induced or augmented labour. Continuous CTG is also recommended when abnormalities are detected during intermittent fetal auscultation.

Good clinical judgement is required to diagnose the underlying cause for a suspicious or pathological CTG, to judge the reversibility of the conditions with which it is associated, and to determine the timing of delivery, with the objective of avoiding prolonged fetal hypoxia/ acidosis, as well as unnecessary obstetric intervention. CTG analysis needs to be integrated with other clinical information for a comprehensive interpretation and adequate management. I hope that many obstetricians can read this book and get familiar with the various CTG interpretations.

I congratulate Dr Roza Olyai on her effort to bring a positive look to women's life through the book edited by her titled *Handbook of Cardiotocography*.

Dr CN Purandare

MD MA Obstetrics (IRL) DGO DFP DOBST RCPI (Dublin)
FRCOG (UK) FRCPI (Ireland) FACOG (USA) FSLCOG (SL)
FAMS FICOG FICMCH PGD MLS (Law)
President, FIGO
Dean, Indian College of Obstetricians and Gynaecologists
President, FOGSI (2009)
President, Indian College of Obstetricians and Gynaecologists (2009)
Professor Emeritus, O&G Research Institute
Ministry of Health, Russian Federation
Editor Emeritus, Journal, FOGSI
Ex. Honorary Professor, Obstetrics and Gynaecology
Grant Medical College and JJ Hospital, Mumbai

Contributors

Achla Batra MD
Professor (Obs and Gyn)
VMMC and Safdarjung Hospital
New Delhi, India

Amanda Henry PhD (UNSW) MPH (UNSW)
B Med Sc (Hons) B Med (Hons) (Newcastle)
FRANZCOG (DDU) (O&G)
Senior Lecturer
Department of Obstetrics
School of Women's and Children's Health
University of New South Wales
Australia

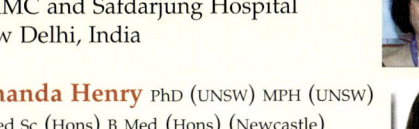

Ameya C Purandare MD DNB FCPS DGO
DFP MNAMS FICMCH FICOG, Fellowship in Gyn
Endoscopy (Germany)
Consultant (Obs and Gyn)
Purandare Hospital, Chowpatty
Mumbai, India

Asanka Jayawardane MBBS MD (O&G)
MPhil (Nottingham) MRCOG
Senior Lecturer and Honorary Consultant
Department of Obstetrics and
Gynecology
University of Colombo/National Hospital
Sri Lanka

Asifa Noreen MD
Senior Registrar (Obs and Gyn)
Services Institute of Medical Sciences
Lahore, Pakistan

Asmita M Rathore MD FRCOG (UK)
Director, Professor
Department of Obstetrics and
Gynecology
Maulana Azad Medical College
New Delhi, India

Bini Ajay FRCOG MSRH FHEA
Consultant (Obs and Gyn)
Croydon University Hospital
London Road, Croydon CR7 7YE, UK

Bushra Haq MD
Senior Registrar (Obs and Gyn)
Services Institute of Medical Sciences
Lahore, Pakistan

Daniella Susic B Med (distinction)
(Newcastle)
Conjoint Associate Lecturer
University of New South Wales Royal
Australian and
New Zealand College of Obstetricians and
Gynaecologists (RANZCOG)
Australia

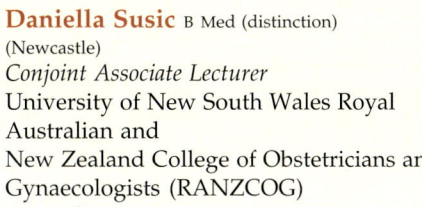

Deepali Mittal Mishra MBBS
Junior Resident (PG 3rd Year)
Department of Obstetrics and
Gynecology
Maulana Azad Medical College
New Delhi, India

Divya Pandey MD
Assistant Professor
Department of Obstetrics and
Gynecology
VMMC and Safdarjung Hospital
New Delhi, India

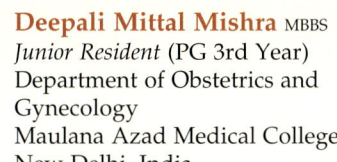
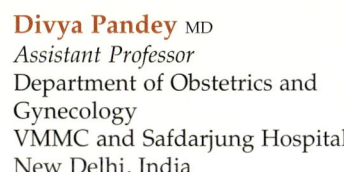
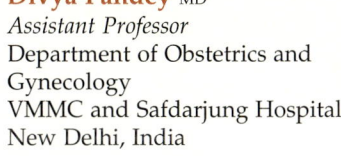

Gita Arjun FACOG
Director
EV Kalyani Medical Foundation
Mylapore, Chennai, Tamil Nadu, India

Madeeha Rashid MD
Assistant Professor (Obs and Gyn)
Services Institute of Medical Sciences
Lahore, Pakistan

Madhuri Mehendale DGO, FCPS, DNB
(O&G)
Assistant Professor
Department of Obstetrics and
Gynaecology
Lokmanya Tilak Municipal Medical College
Sion, Mumbai, India

Maryam Iqbal MD
Assistant Professor (Obs and Gyn)
Services Institute of Medical Sciences
Lahore, Pakistan

Meenoo S
Postgraduate
Maulana Azad Medical College
New Delhi, India

Michelle Mooy MRCOG
Specialist Registrar
Croydon University Hospital
London Road, Croydon CR7 7YE, UK

Mirudhubashini Govindarajan
MBBS FRCSC
Director
Womens Center and Hospitals
Coimbatore, Tamil Nadu, India

Monika Gupta MD
Associate Professor (Obs & Gyn)
VMMC and Safdarjung Hospital
New Delhi, India

Muhunthan K MBBS MS FRCOG
Senior Lecturer
Consultant and Head
Department of Obstetrics and
Gynecology
Faculty of Medicine, University of Jaffna
Sri Lanka

Muralidhar V Pai MS
Professor and Head
Department of Obstetrics and
Gynecology
Kasturba Medical College, Manipal University
Karnataka, India

Nozer Sheriar MD FCPS
Consultant (Obs & Gyn)
Breach Candy, Hinduja Healthcare
Surgical and Holy Family Hospitals,
Mumbai
Aviva Clinic for Women, Mumbai
Past Secretary General, Federation of Obstetric and
Gynecological Societies of India
Past President, Mumbai Obstetric and Gynecological
Society, India

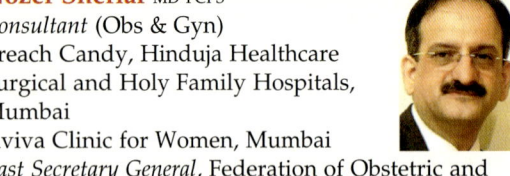

Nuzhat Aziz DNB DGO (O&G)
Consultant and Head
Department of Obstetrics
Fernandez Hospital,
Hyderabad, Telangana, India

Pallavi Chandra RMS (O&G)
Consultant
Department of Obstetrics
Fernandez Hospital
Hyderabad, Telangana, India

Parikshit D Tank MD DNB FCPS DGO DFP
MNAMS MICOG FRCOG
Consultant
Ashwini Maternity and Surgical
Hospital, Mumbai
Consultant
Jupiter Hospital, Thane
Maharashtra, India

Pratima Mittal MD
Professor and Head
Department of Obstetrics and
Gynecology
VMMC and Safdarjung Hospital
New Delhi, India

Rachael Yates
CHSALHN *Midwife Manager*
Maternal and Neonatal Services
Country Health SA Local Health
Network
SA Health Government of South Australia
South Australia

Rajneet Bhatia MD
Clinical Associate (Obs and Gyn)
Hinduja Healthcare Surgical Hospital
Mumbai, Maharashtra
Aviva Clinic for Women
Mumbai, Maharashtra, India

Rekha Upadhya MBBS DNB
Assistant Professor
Department of Obstetrics and
Gynecology
Kasturba Medical College
Manipal University
Manipal, Karnataka, India

Reva Tripathi MD
Professor and Head
Department of Obstetrics and
Gynecology
Jamia Hamdard University
New Delhi, India

Rohana Haththotuwa MBBS MS (O&G)
FICOG (Hon) FSLCOG FRCOG
Consultant (Obs and Gyn)
Chairman
Ninewells Care Hospital
Colombo, Sri Lanka

Rubina Sohail MD
Professor (Obs and Gyn)
Services Institute of Medical Sciences
Lahore, Pakistan

Professor Sir Sabaratnam Arulkumaran MD PhD
FRCS FRCOG
Emeritus Professor (Obs and Gyn)
St George's University of London, UK
Foundation Professor (Obs and Gyn)
University of Nicosia, Cyprus and
Visiting Professor
Institute of Global Health
Imperial College London, UK

Sameer Dikshit FRANZCOG MD DGO
FCPS FICOG
Consultant (Obs and Gyn)
Mount Gambier and Districts Hospital
Mount Gambier, South Australia
Lecturer
Flinders' Rural Medical School
Flinders' University
South Australia

G Selvanandhini Gopalasundaram
MD Fellowship in Fetal Medicine
Consultant
Maternal Fetal Medicine
Womens Center and Hospital
Coimbatore, Tamil Nadu, India

Sonali Gaur MD FRCOG
Consultant (Obs & Gyn)
Nanavati Superspeciality Hospital
Mumbai, Maharashtra

Visiting Consultant
Hinduja Healthcare, Khar and
Surya Mother and Child Care
Mumbai, Maharashtra, India

Sridevi Veluganti Nagasai
DNB (O&G)
Senior Registrar
Department of Obstetrics
Fernandez Hospital
Hyderabad, Telangana, India

Vidyashree Poojary MS DNB
Assistant Professor
Department of Obstetrics and
Gynecology
Kasturba Medical College
Manipal University
Karnataka, India

Vikram Sinai Talaulikar MD
MRCOG PhD
Associate Specialist
Reproductive Medicine Unit
University College London Hospital
London, UK

Preface

It gives me great pleasure to share with you the book titled *"Handbook of Cardiotocography"*

Intrapartum perinatal mortality is a major issue in the developing countries which accounts for more than one million stillbirths worldwide each year. In order to decrease the rate of stillbirths and neonatal mortality, fetal monitoring is essential. This book deals with the CTG monitoring, starting from how to interpret a normal CTG to the various abnormal traces, its necessity in high-risk patients, medicolegal aspects, the latest recommendations and NICE guidelines.

In developing countries like India, where the doctor to patient ratio is low, one to one monitoring is not possible for every patient either by a doctor or by a midwife. CTG monitoring has become an important tool in an obstetrician's life both in the government and the corporate sector. Hence, this book will help the budding obstetricians, junior residents, interns, students and midwives to have a clear concept about various methods of intrapartum fetal monitoring and the recent advances in this field. Even though this book mainly focuses on CTG, we have covered other methods of monitoring like use of fetal scalp monitor, fetal blood sampling, fetal ECG and fetal pulse oximetry.

Misinterpretation of a normal CTG like wrong classification of decelerations may lead to unnecessary obstetrical interventions, instrumental deliveries and caesarean sections. Therefore, it is important to know the basic features and analysis of CTG. It is essential for an obstetrician to differentiate between a fetal stress pattern which is normal during labour and a fetal distress which may require an urgent intervention.

After reading this book, one can have better understanding regarding the various CTG readings, identifying abnormal traces, resuscitating the patient and managing the situation by taking proper actions for delivering the fetus in case of fetal decompensation.

Considering the increasing awareness of healthcare amongst the public during the last two decades and the various medicolegal litigations that are raised against various doctors, documentation is very important on our side. Hence CTG method of monitoring would help an obstetrician to survive any litigation against him/her in case of any mishap if occurred to the mother or baby despite proper actions taken.

This book contains chapters contributed by experienced faculties in the field of obstetrics and also has samples of various CTG traces that may help the readers for a better knowledge and perception of the subject.

Wishing you a happy reading!

Roza Olyai

rozaolyai@gmail.com

Acknowledgments

I am honoured that Professor (Dr) CN Purandare, President, International Federation of Gynecology and Obstetrics (FIGO), has graciously concurred to write the foreword.

I would like to sincerely thank Sir Sabaratnam Arulkumaran, Professor Emeritus of Obstetrics and Gynaecology, St George's University of London, for his support and encouragement throughout the journey for this book. It was due to his guidance that I could put these chapters together and gather such good galaxy of authors. I would also like to thank Dr Shirish Daftary, Professor Emeritus of Obstetrics and Gynaecology and Past President of the Federation of Obstertics and Gynecological Societies of India who has been my mentor and guide throughout my career.

My special gratitude to Mr Johan Vos, Chief Executive, International Federation of Gynecology and Obstetrics for his words of encouragement sharing his views with the readers.

I would like to thank all my contributors from India and abroad of various specialties and organizations for their generous contribution towards this book. I am grateful to Mr SK Jain and the staff of CBS Publishers & Distributors Pvt. Ltd. for publishing this book and help spread the message, in particular Mr Ramesh Krishnamachari for his good coordination. Last but not least, I am greatly indebted to my parents who have given me the courage to face challenges of the world by inculcating good attributes in me, to my loving and supportive husband Dr Nayson Olyai, for his constant support and being my best critic, my son Navid, my daughter Neda, my son-in-law Vahid and my granddaughter Norah, for their tolerance and understanding of the time away from them during my career and in editing this book and being the backbone to my progress.

Contents

Cardiotocography Today

Nuzhat Aziz

Cardiotocography (CTG) is the most commonly used procedure in labour.[1] High income countries have a higher rate of women undergoing electronic fetal monitoring when compared to low resource countries. Evaluating, discussing the current status of cardiotocography would require a look back on the evolution of fetal monitoring and the reasons for its status today.

There was a time when a baby's cry was the only sign of fetal well-being at birth, until a startling revelation from Marsac in 1650 that we could hear the heart beat of the fetus like 'beating like the clapper of a mill'. The first recorded literature for Fetal Heart Rate (FHR) detection is credited to Kergaradec in year 1818 and since then we have had all research targeted towards better identification of the fetal heart sounds, for over 100 years. From the detection of an alive fetus, progressive efforts were directed at identification of a sick, hypoxic fetus through the FHR characteristics. In the year 1833, Evory Kennedy recommended hearing the FHR as a means of monitoring the fetus in labour. The criteria for fetal distress was established by von Winkel in 1893 and correlated to the adverse outcomes at birth.[2] In 1906, Max Cremer developed a galvanometer to record the electrical fetal activity for the first time in Germany, indirectly through maternal limb leads. The fetoscope developed by David Hills and Joseph DeLee in 1917 was widely accepted and maintained its status as the best

fetal monitoring tool for long-time. The work on the capture of the fetal electrocardiograph (ECG) continued and George H Bell could obtain a direct fetal ECG using two metal electrodes on the uterus in 1938. At about this time, Edward Hon was working on amnioscopy to detect meconium staining of amniotic fluid as a marker for a sick fetus. He then moved to assessing the fetal pH as a marker for acidosis and developed an electronic monitor which could record the FHR continuously in 1958. Hon in America, Caldeyro-Barcia in Uruguay and Hammacher in Germany are credited for their pioneering work in the development of electronic fetal monitoring equipment, which was commercially available in 1962.

Kubli and his colleagues reported their research on fetal pH and the FHR patterns in the year 1969.[3] This was a breakthrough finding in the history of fetal monitoring; stimulated many to think about the possibility of continuous capturing of the FHR patterns. The hope was to eliminate all intrapartum hypoxia related events.

The first study to correlate the FHR patterns with fetal pH was done on eighty five women and published in the year 1969.[3] Severe variable and late decelerations were most commonly associated with a pH of less than 7.25. They reported that the most important finding of their study was a good correlation between a normal trace and absence of low pH. In the 1970s and 1980s, rhesus monkey based studies provided new knowledge about the FHR patterns in hypoxia, including those preceding fetal death. Decrease in uterine blood flow and oxygenation was associated with late decelerations and low fetal pH was a very late finding. Persistant late decelerations without accelerations was the terminal pattern preceding death.[4]

Low et al. compared traces of 71 babies born with metabolic acidosis (with base deficit of more than 16) with 71 controls without acidosis. They found 68% had decreased or minimal baseline FHR variability in acidotic group. Four babies in the study group had no CTG evidence of hypoxic changes. Decreased variability with decelerations was the most predictive pattern for neurological injury to the fetus.[5]

A randomised control trial designed to evaluate the efficacy of Electronic Fetal Monitoring (EFM) over intermittent auscultation was stopped as interim analysis showed significantly lower Neonatal Intensive Care Unite (NICU) admissions in the EFM group.[6] This was followed up by many randomised controlled trials with different study groups; unselected vs selected, low-risk vs high-risk. Few studies had a decrease in perinatal deaths, but majority reported a decrease in only two outcomes; neonatal admissions and neurological symptoms.

In 1980s, efforts were targetted to study of benefits in highly selected population, e.g. preterm. The consistent findings were the same; the benefits from electronic fetal monitoring did not match the expectations. The associated increased cesarean section rate with EFM became a concern, confirming a high false positive rate for prediction of fetal hypoxia. Cochrane Database reviewed and confirmed these findings. Electronic fetal monitoring in labour was associated with increased cesarean section and operative vaginal delivery rates.[7]

The difficulties in interpretation of these studies were due to a wide variation in definitions, classification and terminology. The

Eunice Kennedy Shriver National Institute of Child Health and Human Development (NICHHD) formed a collaborative group and issued standardised definitions, followed by a three tier classification in 1997 and 1999. The National Institute of Clinical Excellence (NICE), the American College of Obstetrics and Gynecology (ACOG) and the International Federation of Obstetrics and Gynecology (FIGO) classification are the most widely used and differ slightly in the definitions, terminology and management algorithms.[8–11]

How do we decrease the false positive rates? A confirmatory test with fetal blood sampling seems to be the best available tool currently. The countries who have used this step in their management algorithms, as a confirmatory test have maintained their cesarean section rates around 25%. The efficacy of fetal blood sampling, scalp pH for predicting a hypoxic baby has been questioned by many.

Today we are aware of no advantages of electronic fetal monitoring in low-risk women; the best tool appears to be intermittent auscultation with least intervention rates. The high-risk women have the recommendation of cardiotocography as the basic monitoring tool, but studies have not able to demostrate a statistically significant reduction in hypoxic perinatal deaths and morbidity. Fetal pulse oximetry, infrared spectometry, segment (ST) waveform analysis (STAN) and many others were tried as better tools. None were found to be good predictors of a sick hypoxic baby; they were good at predicting a healthy baby.

The rates of intrapartum stillbirth and neonatal encephalopathy vary across the globe; from one country to another; from one institution to the other.[12] The confidential enquiries into stillbirths highlighted the importance of many aspects of cardiotocography beyond a capture of a continuous FHR trace. The interpretation of CTG, classifications and algorithms came under scrutiny. The most common allegation in a

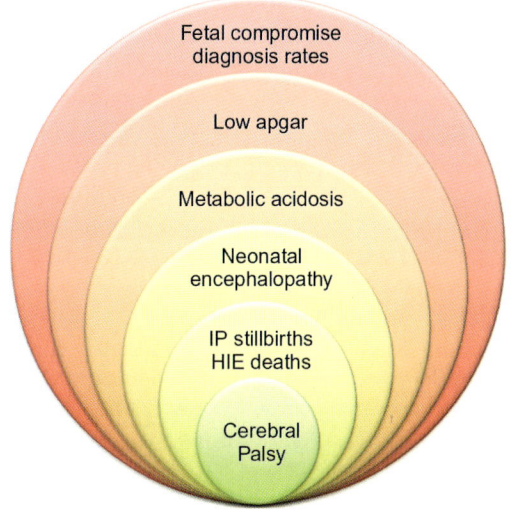

Fig. 1.1: Audit parameters.

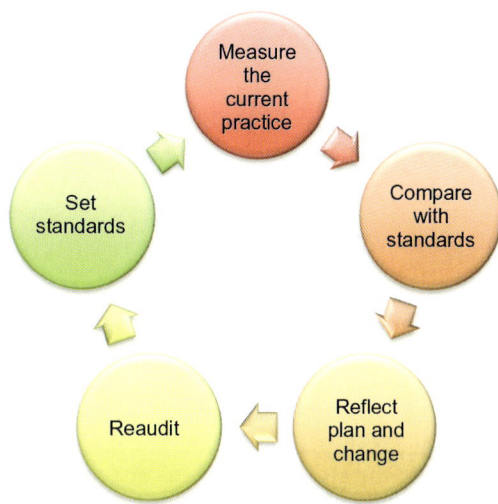

Fig. 1.2: Audit cycle.

medicolegal evaluation was delayed diagnosis and inappropriate response. The softer skills of commmunication, escalation to a proper chain of command, instituting the most appropriate response to an abnormal trace were identified as areas to train and improve. The rates of asphyxia related morbidity and mortality have been to shown to decline over past two decades. A population based retrospective study from a registry of births for Scotland from 1988 to 2007 showed a significant decrease in intrapartum stillbirths by 38%. The decrease was mainly in the risk of death ascribed to intrapartum anoxia (5.7 to 3.0 per 10,000 births; unadjusted change, –48%; 95% CI, –62% to –29%).[13] Intrapartum care, management policies with an emphasis on team work have been a focus of improvement in past two decades. Cardiotocography has a definite role in these improved outcomes but not the only tool.

Cardiotocography in India is at a phase of introduction of machine to capture FHR currently. The worry of increased interventions is beginning to show with rapid increase in cesarean section rates across all states. India needs guidelines for fetal monitoring, training modules, certification and standardization of classifications. Defining and auditing perinatal outcomes, identifying gaps, setting up goals to reduce the disparity in outcomes between different setups should be the first step (Figs 1.1 and 1.2). CTG is a procedure which has proved itself as a good screening tool for absence of hypoxia. CTG interpretation and management is a skill which needs to be learnt, with an agreed curriculum, periodically updated and certified before they take charge of intrapartum care.[14]

REFERENCES

1. Parer JT. Electronic fetal heart rate monitoring: a story of survival. Obstet Gynecol Surv 2003; 58:561–3.

2. Whitfield CR. Foetal heart rate monitoring—present lessons and future developments. The Ulster Medical Journal. 1966;35(1):75–82.

3. Kubli FW, Hon EH, Khazin AF, et al. Observations on heart rate and pH in the human fetus during labor. Am J Obstet Gynecol 1969;104(8): 1190–206.

4. Myers RE, Mueller-Heubach E, Adamsons K. Predictability of the state of fetal oxygenation from a quantitative analysis of the components of late deceleration. Am J Obstet Gynecol 1973; 115(8):1083–94.

5. Low JA, Victory R, Derrick EJ. Predictive value of electronic fetal monitoring for intrapartum fetal asphyxia with metabolic acidosis. Obstet Gynecol 1999;93(2):285–91.

6. Renou P, Chang A, Anderson I, et al. Controlled trial of fetal intensive care. Am J Obstet Gynecol 1976;126(4):470–6.

7. Alfirevic Z, Devane D, Gyte GML, Cuthbert A. Continuous cardiotocography (CTG) as a form of electronic fetal monitoring (EFM) for fetal assessment during labour. Cochrane Database of Systematic Reviews 2017, Issue 2. Art. No.: CD006066. DOI: 10.1002/14651858.CD006066.pub3

8. National Collaborating Centre for Women's and Children's Health commissioned by the National Institute for Health and Clinical Excellence. Intrapartum care. 2007.

9. American College of Obstetricians and Gynecologists. ACOG Practice Bulletin No. 106: Intrapartum Fetal Heart Rate Monitoring: Nomenclature, Interpretation, and General Management Principles. 2010 Compendium of Selected Publications, 324–34.

10. American College of Obstetricians and Gynecologists. ACOG Practice Bulletin No. 116: Management of Intrapartum Fetal Heart Rate Tracings." Obstet Gynecol 2010;116(5):1232–40.

11. Ayres-de-Campos D, Spong CY, Chandraharan E, for the FIGO Intrapartum Fetal Monitoring Expert Consensus Panel. FIGO consensus guidelines on intrapartum fetal monitoring: Cardiotocography. Int J Gynecol Obstet 2015; 131:13–24.

12. Goldenberg RL, McClure EM, Bann CM. The relationship of intrapartum and antepartum stillbirth rates to measures of obstetric care in developed and developing countries. Acta Obstet Gynecol Scand 2007;86:1303–9.

13. Pasupathy D, Wood AM, Pell JP, Fleming M, Smith GC. Rates of and factors associated with delivery-related perinatal death among term infants in Scotland. JAMA2009;302:660–8.

14. Ugwumadu A, Steer P, Parer B, Carbone B, Vayssiere C, Maso G, Arulkumaran S. Time to optimise and enforce training in interpretation of intrapartum cardiotocograph. BJOG 2016; 123:866–9.

The CTG Machine

Nuzhat Aziz, Sridevi Veluganti Nagasai

Cardiotocography (CTG) machines are advanced complex electronic machines which are capable of acquiring, processing, displaying and printing the Fetal Heart Rate (FHR) patterns and uterine contractions in pregnancy. Hon, Caldeyro-Barcia and Hammacher are credited for developing the technique of CTG in the 1960s in different countries, which was made available for commercial use in 1967.[1] This chapter is aimed at gaining more knowledge of the monitors, settings and technical difficulties, helping in optimal usage of the machine for a test that has become the most frequently used obstetric procedure.

Functions of a CTG Machine

The CTG machine has the basic function of recording the FHR trace and uterine contractions against time. The machine has an electronic display for the recorded para-meters, speakers for an audio output and a thermal printer to give a printed output simul-taneously. The fetal movement recording has been added as a basic function, as perceived by the woman and the ultrasound FHR transducer. The fetal monitoring strips are a legal record and should be compatible with the manufacturer and instruments specifica-tions. The paper speed can be changed through settings depending on the user preference.

The Components of a CTG Machine

The parts of a CTG machine are shown in Fig. 2.1 with a difference in the special adapters and tools across different models and manufacturers. The following tools are available with a machine:
- Toco transducer (with or without maternal pulse)
- US transducer (ultrasound)
- External power supply cord
- Event marker (maternal fetal movements perceived)
- Fetal paper pack
- Belts with button or Velcro fixings
- DECG adapter cable (Direct ECG)
- MECG adapter cable (Maternal ECG)
- IUP adapter cable

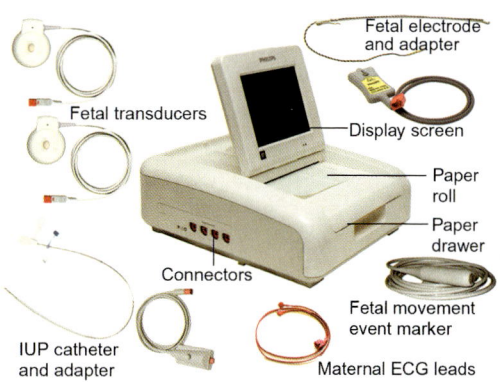

Fig. 2.1: Components of a CTG machine.

The internal monitoring requires disposable catheters, which are sterile, preassembled and ready to use and are available in India.

External Cardiotocography

External cardiotocography has an ultrasound based principle (Doppler) for detection of cardiac movement and then converts the signal to a line trace with signal modulation. Real-time ultrasound Doppler technique was developed in 1964. It is used to obtain the FHR from the movement of the cardiac walls, valves, which is picked up as reflected signals by the transducer. The placement of peizo-electric crystals are different in continuous and pulsed Doppler techniques. The transducers having pulsed Doppler technique have only one crystal which emits and then receives the reflected signals. The transducers with continuous Doppler technology have a central transmitting crystal and reflected signals are received by six surrounding elements. The pulsed Doppler technique has the advantage of exposing the fetus to lower sound energy. The commonly used transducers use an ultrasound frequency of 1 MHz $\pm$ 100 Hz.

In the first generation CTG machines, the signal modulation was not very well developed, accounting for a lot of signal noise. The signal modulation improved significantly in the second generation machines which have the capability to modulate the signal, have spike removal and autocorrelation to give a much better trace than before. Autocorrelation used the inter beat interval to calculate a rate and does not precisely reflect the fetal heart beats. This was made possible through simultaneous development of technology through microprocessors.[2-5]

The signal received by the transducer can be affected by movement of any intra-abdominal structures, e.g. blood vessels, intestinal and fetal movements, resulting in recording of mixed signals. A CTG trace can only be obtained if the signal failure is less than 15% of the time. Loss of signals can also be due to inadvertent pick up of maternal pulsations or due to technical machine error of capture of half the cardiac beats or doubling of the beats. The users must be aware of the technical problem of halving or doubling of the beats which may be a source of error in interpretation of the trace. The signal capture becomes difficult for pregnancies with more than one fetus–multifetal pregnancies. The second generation machines with dual monitoring channels for twin pregnancies have the capability to identify if the signals are similar and are able to give an alert for it. The fetal transducers should be applied after proper localisation of the fetal heart, well differentiated from the maternal pulsations. External method of cardiotocography is the routine recommended method for ante-partum and intrapartum fetal surveillance.

The signal autocorrelation has a drawback for its inability to compute a trace in cardiac arrhythmias. Arrhythmias are difficult to capture with external cardiotocography and may require either an ultrasound to confirm or an internal electrode to record. Contra-indications to external cardiotocography are magnetic resonance imaging, defibrillation and electrosurgery. The transducer has been made waterproof but care should be taken while using electrical equipment under water.

Internal Cardiotocography

Internal monitoring requires the application of an electrode to the fetal presenting part, usually the scalp. It is only an intrapartum tool and can be applied after 2 to 3 cm of cervical dilatation with rupture of amniotic membranes. The technology used for capturing the FHR trace is very different from that of external cardiotocography. The electrode captures the fetal electrical cardiac activity and gives a FHR trace from the R – R interval, which has much lower signal loss rate. This technology is the most reliable way of obtaining a FHR trace. The electrode may sometimes give the maternal signals if it is applied onto the cervix or in cases of intrauterine fetal death when there are no fetal electrical signals. Contraindications to the internal monitoring are extreme prematurity (less than 32 weeks), intact membranes, non-vertex presenting part and conditions posing risk of transmission of infection to the fetus. Fetal electrode is a thin spiral metal electrode also called screw or scalp electrode as shown in Fig. 2.2. There are other types of electrodes available, with clips, with different mechanisms to prevent dislodgement. The spiral electrode comes with a protective sleeve and can be screwed onto the scalp easily during a vaginal examination. It has its advantages when the external probes are not able to pick up the signals very well, e.g. in morbid obesity, in multifetal pregnancy or in second stage of labour.

The internal cardiotocography is an optional tool in the CTG machine and is available in selected models. It requires a connector cable which is fixed onto the woman's thigh, then onto the machine in any of the fetal transducer slots. The machine automatically switches over to fetal electrode based signals which are displayed in the monitor in a different colour. To maintain uniformity, application is done by clockwise screwing and removal by anticlockwise unscrewing. The electrode is removed by unscrewing in the opposite direction.

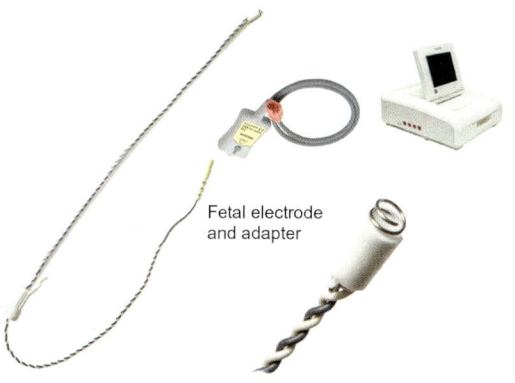

Fig. 2.2: Fetal scalp electrode.

Fetal electrode and adapter

Removal of the electrode can be after complete delivery of the baby or at crowning. It may have to be removed prior to cesarean section or in cases of wrong application. The wound created by the fine thin electrode heals rapidly. Internal cardiotocography is invasive, costlier than external methods and used only in special cases.

MONITORING OF UTERINE CONTRACTIONS

Tocography is recording of uterine contractions over time. It can be done via the external transducer or through an Intrauterine Pressure Catheter (IUPC). The external method is a pressure sensitive tocodynamometer which detects the upward uterine movement associated with a contraction and gives a bell shaped trace. The toco transducer is held in place by belts on the maternal abdomen at the level of the fundus. The placement and sensitivity of the knob determines the width and the height of the uterine curve. A highly sensitive transducer is capable of detecting minimal pressure changes and creates a wider curve over a longer period of time. The tightness of the belts determine the amplitude of the curves. External tocography can detect only the frequency of the contractions and is not capable of detecting other characteristics; basal tone, amplitude and duration of contractions. Baseline setting is a vital part of obtaining a good contractions trace. This setting, also called uterine activity reference, has to be used during a uterine relaxation phase to bring the trace to a predetermined baseline. This is set at 20 or 10 mm Hg and does not denote the intrauterine pressure of the uterus during relaxed phase—the baseline tone. The signal quality of the uterine contractions trace can be affected by improper placement, tight placement, adiposity and loss of sensitivity of the transducer.

Disadvantages of external tocography are failure to determine the amplitude or strength of contractions, difficulty in obtaining a trace in the lateral positions and signal interference with maternal movements. External tocography is the most widely used and recommended method for monitoring uterine contractions.

INTERNAL-INTRAUTERINE PRESSURE CATHETER

The first Intrauterine Pressure (IUP) catheter was developed in the 1950s. This device is placed inside the amniotic cavity to measure intrauterine pressure, hence quantifying the strength of uterine contractions (in Montevideo units). The catheter was originally an open tube pressure monitoring tool; evolving to a small transducer at the tip of a pressure catheter. The current catheters have a small catheter tip pressure transducer, which records the pressure. It has multiple advantages over the fluid filled IUP catheters and are more accurate. Some catheters have a port available for amnioinfusion.

The IUP catheters can be inserted only in labour after amniotic membrane rupture. The use of IUP has declined over time with very few recommending it today. It has similar risks as fetal electrode for infections but rare additional risks of placental abruption, fetal injury, displacement and rarely uterine rupture have been described. The contraindications to IUP placement are antepartum hemorrhage and placenta previa. Intrauterine pressure catheter is not recommended for a routine use and has not been found to be helpful. It has not been found to be of benefit in induced or augmented labour too. Both fetal electrode and intrauterine pressure catheters are available in India, but require a compatible machine with facility to use these.

RECORDING OF FETAL MOVEMENTS

Recording fetal movements in a CTG trace is helpful for antenatal fetal surveillance. It gives added information when a Nonstress Test (NST) is requested for complaints of decreased fetal movements. It can be done by an event marker which is given to the mother to press when she perceives a movement. Few machines have ability to pick up low-frequency fetal movements from the ultrasound transducer. The transducer seives out the low-frequency fetal movements from the high-frequency cardiac movements and gives a recording of the fetal activity over a period of time (Fig. 2.3). It is been designated by some as actocardiography and has a drawback that it cannot differentiate active from passive fetal movements.[6] The fetal movements which are in the line of the ultrasound beam are detected, and hence may not be a reliable indicator of absence of fetal movements. The fetal movement markings sometimes help in determination of a baseline. Some traces give the Fetal Movements Perceived (FMP) as the percentage of time and represents on the trace as "FMP".[7]

MATERNAL HEART RATE MONITORING

The maternal heart rate trace should become an integral part of all machines. Often the maternal heart rate rises due to stress of labour, and sometimes leads to interpretation errors. Maternal heart rate with accelerations may be misinterpreted for a normal FHR trace with accelerations. Critically ill acidotic women with tachycardia may have a hypoxic fetus. Intrauterine fetal deaths have sometimes been misinterpreted as normal traces, which were maternal pulsations instead recorded by the transducer. The users should be aware of this possibility. Shift of signal to maternal pulsations may mimic a deceleration leading to unnecessary intervention.

Maternal aortic pulsations are recorded by a small ultrasound transducer placed in the tocodynamometer. Few manufacturers upgrade the machines to include this facility on request. If the maternal ECG leads have been used this maternal pulse rate trace is taken from more reliable maternal ECG signals which are displayed in the monitor with a different colour.

Additional Tools

The current machines have optional maternal ECG, blood pressure, pulse oximetry incorporated. They have the advantage of reducing the number of machines/gadgets for a single woman who requires intensive maternal and fetal monitoring. The disadvantage is cost. The maternal ECG is a very useful tool in obese women where there is loss of signal from the fetus as well as the maternal pulse rate. Computerised decision-making support systems are being increasingly added in newer models, which have not been found to have any improvements in outcomes.

Machine specifications[7]

1. **Ultrasound transducer**
   - Ultrasound frequency used 1 MHz $\pm$ 100 Hz
   - Method: Ultrasound pulse Doppler
   - Ultrasound intensity: Average output power 7.4 $\pm$ 0.4 mW
   - Ultrasound burst: Repetition rate 3.0 kH, with duration of $\leq$100 ms
2. **Toco transducer**
   - Range 40–240 bpm for maternal pulse
   - Resolution 1 bpm
   - Accuracy $\pm$ 2% or 1 bpm whichever is greater
3. **ECG:** DECG and MECG both have measurement range of 30–240 bpm with an accuracy of $\pm$ 1 bpm or 1%, whichever is greater.
4. **IUP:** Though used less often in clinical practice, it has a measurement range of –100 to +300 mm Hg, accuracy of $\pm$ 0.5% per 100 mm Hg.

MACHINE SETTINGS

The settings on CTG machines should be standardised, in accordance with the recommendations. The users need to understand them to ensure optimal use of the machine. The mandatory settings are paper speed, date and time, which should be locked and not allowed to be edited, for the trace is a medicolegal document and loses its value if the name, identification number, date and time are not entered. The user manual has all the information on how to change the settings, and should be requested at the purchase of every machine. For twins and triplets CTG monitoring a trace separator is present. For better visualisation of the traces and to separate them from each other, the FHR2 channel is separated from FHR1 by +20 bpm. This means that, e.g. if FHR2 shows 160 bpm, it is actually 140 bpm. Machine are available for triple channel monitoring for triplets where FHR2 is +20 bpm and FHR3 is –20 bpm.[7]

Alerts

The audio alerts can be set manually for the FHR with a minimum and maximum. There is a cross channel verification[7] between the two ultrasound transducers for FHR and the maternal pulse rate, which can give alerts on the display monitor and on the printed trace (Fig. 2.3).

RECORDINGS ON A CTG MACHINE DISPLAY AND TRACE[7]

- Patient's demographics like name, age, identification number, last menstrual period, expected date of delivery.
- Fetal heart rate
- Uterine activity
- Fetal movement profile
- Maternal pulse using the same external toco transducer
- Maternal heart rate via maternal ECG electrodes
- Non-invasive blood pressure
- Additional facilities are internal fetal monitoring via fetal electrode, uterine activity with IUP catheter and maternal oxygen saturation (SpO_2).

- Pre-configured notes can be entered such as lateral position, vaginal examination, rupture of membranes, etc. which are then printed on the trace.

Central Monitoring Station-digital Output and Data Storage

Systems are available for integrating all the machines in a unit to a central monitoring station, with settings for individual machine display or a combined display. There are digital storage systems available to store the traces electronically and retrieve when needed. Cardiotocographs should be stored for a minimum of 25 years, and are known to fade away over a period of time (thermal paper). They can be stored for a longer time by protecting from light in thick brown envelopes. A photocopy of these traces can be considered when there are adverse neonatal outcomes and a developmental delay is suspected. Tracer systems should be present if they are stored separately from her complete hospitalization records.

Telemetry

Restriction of movement is one of the major disadvantages of being connected to a CTG machine. This disadvantage was removed by use of wireless probes, which can transmit signals within a specified distance.[8] Range varies depending on the manufacturer, but are a useful addition. These probes are also waterproof and can be used under water, if needed.

CTG Trace Markings

Identification of all the markings on the CTG trace allows for better understanding of the traces. Figure 2.3 shows the different parts of a CTG trace.

A. **Identification:** Name, ID, date, time, the ID of the different probes gets printed at the beginning of the trace. These are of legal implications for a CTG can be traced to a respective machine and probe. A CTG without these details loses its medicolegal value.

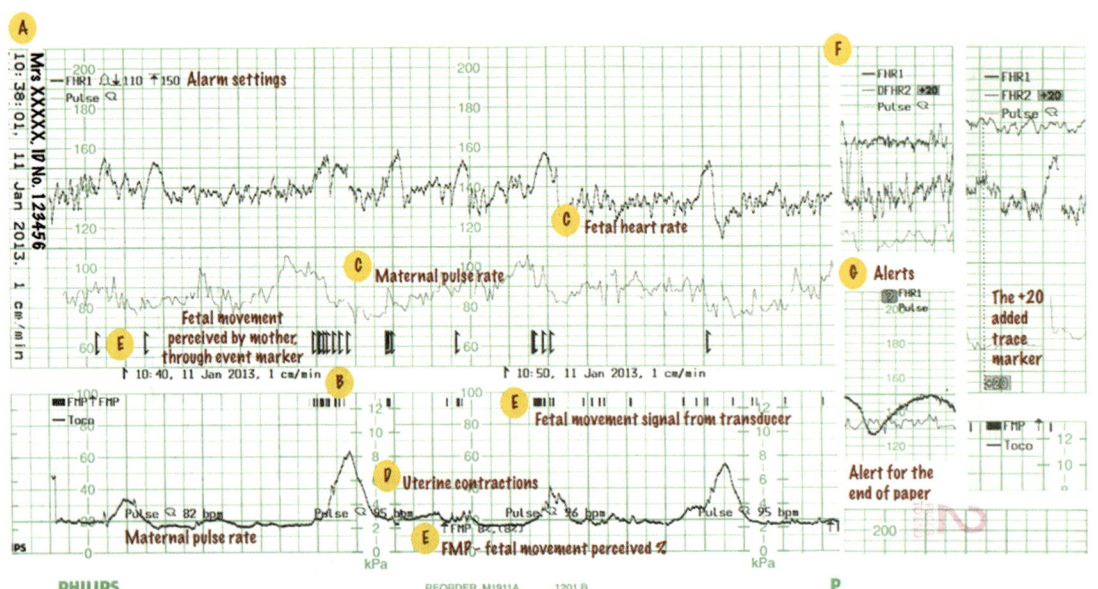

Fig. 2.3: Reading a CTG.

B. **Print paper settings:** The recommendations are given for the CTG paper but the one most familiar to the team can be used. The paper speed determines the time interval; 1 cm/minute speed would lead 1 cm representing 60 seconds and 0.5 cm representing 30 seconds and so on. The National Institute for Clinical Excellence and International Federation of Obstetrics and Gynecology guidelines recommend the following settings.[8–10]

- Paper speed is set to 1 cm/minute
- Sensitivity displays are set to 20 bpm/cm
- FHR range displays of 50–210 bpm are used.

The standard markings of date, time, paper speed is printed every 10 minutes. The preprinted markings on the paper (scale) is usually at 10 cm distance. A 10 cm of paper represents 10 minutes with a paper speed of 1 cm/minute, likewise a paper speed of 3 cm/minute leads to 30 cm of paper representing 10 minutes.

C. **The traces:** The FHR and maternal pulse rate traces are as shown in Fig. 2.3. The FHR is a dark line when compared to the maternal pulse. The scale for heart rate traces is different in different machines. Interpretation difficulties occur when there is a sudden change in the scale or speed not familiar to the team.

D. **The uterine contractions:** The scale for uterine contractions is mm Hg, but this is of relevance only when an Intrauterine Pressure (IUP) catheter is used. The uterine activity can be measured in montevideo units or uterine activity integers only if an IUP is placed. The external tocography gives only the frequency of contractions which is represented as number of contractions in 10 minutes.

E. **Fetal movement perceived:** The event marker action is printed on the trace as an arrow marking a Fetal Movement Perceived (FMP) by the woman. The ultrasound transducer identified fetal movements is printed on a continuous bar with print representing movement. The percentage of time is represented as FMP which is through the ultrasound transducer.

F. **Twin pregnancy markers:** The use of two external FHR probes is denoted on the trace as FHR1 and FHR2. The use of fetal electrode is denoted by DFHR (direct FHR) and IUP as IUP. The traces have different grey scales and weight to differentiate betweem them,

G. **Alerts:** Queries through cross channel verification system when there is synchrony in rates through two transducers. The paper strip has preprinted alerts to mark the ending of the paper roll.

CONCLUSION

Cardiotocography has become an integral part of day-to-day obstetrics. Knowledge about a machine and its functions helps in better utilisation and prevention of inadvertent errors. External cardiotocography is the most commonly used and recommended modality with internal methods being reserved for special situations.

REFERENCES

1. Jauniaux E, Prefumo F. Fetal heart monitoring in labour: from pinard to artificial intelligence. BJOG 2016;123(6):870. doi: 10.1111/1471–0528.13844.
2. Gunn AL, Wood MC. The Amplification and Recording of Foetal Heart Sounds [Abridged]. Proceedings of the Royal Society of Medicine 1953;46(2):85–91.
3. Horton R. The significance of autocorrelation in fetal monitoring. Midwives Chron 1991; 104(1244):260–1.
4. Carter MC. Signal processing and display—cardiotocographs. Br J Obstet Gynaecol. 1993; 100(Suppl 9):21–3.
5. Peters M, Crowe J, Piéri J, et al. Monitoring the fetal heart non-invasively: a review of methods. Journal of Perinatal Medicine 2005;29(5):408–16.

Retrieved 10 Sep. 2017, from doi:10.1515/JPM. 2001.057

6. Maeda K. Invention of ultrasonic Doppler fetal actocardiograph and continuous recording of fetal movements. J Obstet Gyanaecol Res 2016; 42(1):5–10. doi: 10.1111/jog.12855. Epub 2015.

7. Avalon fetal monitor FM20/30, FM40/50 by Philips Training Guide, Release G.0 with software revision G.02.xx; 2010, printed in Germany.

8. Ayres-de-Campos D, Nogueira-Reis Z. Technical characteristics of current cardiotocographic monitors, Best Practice and Research Clinical Obstetrics and Gynaecology, doi.org/10.1016/j.bpobgyn.2015.05.005.

9. National Collaborating Centre for Women's and Children's Health commissioned by the National Institute for Health and Clinical Excellence. Intrapartum care 2007.

10. Ayres-de-Campos D, Spong CY, Chandraharan E, for the FIGO Intrapartum Fetal Monitoring Expert Consensus Panel. FIGO consensus guidelines on intrapartum fetal monitoring: Cardiotocography Int J Gynecol Obstet 2015;131: 13–24.

Indications for CTG Monitoring

Parikshit D Tank

Cardiotocography (CTG) is a simple, low-resource, noninvasive test of fetal well-being. Like all tests of fetal well-being, the rationale of CTG is to identify the fetus with hypoxia so that intervention (usually delivery) is well timed and further fetal damage is avoided. However, the inherent flaw in CTG is the high false positive rates which could lead to unnecessary interventions.[1] Therefore, it is essential to select a population of pregnancies which have a risk of fetal hypoxia and improve the performance of the test. Since its inception into clinical practice in the late 1970s, the CTG has become a popular test. It has no known absolute contraindications. The indications for CTG in the antenatal and in labour are debatable. This chapter focuses on indications and will cover brief reviews, rationales and evidence in specific clinical situations.

Period of Gestation

All the key features of a CTG which indicate a normal fetal status are altered in preterm gestations. This is due to the yet-to-mature autonomous nervous system. The parasympathetic centers mature later and the heart rate and its patterns are largely sympathetically driven. Also, the balance between the two is not well-established. The baseline heart rate is higher, there is less variability and there are no or few accelerations in fetuses remote from term. This results in more traces being labeled nonreassuring or abnormal. The test has an even higher rate of false positive results.[2] This could result in interventions and/or delivery far removed from term, making it a potentially harmful test to use in preterm gestations. Most units do not perform CTG before 32 weeks of gestation. In units doing CTG at earlier gestations, the interpretation and further action on the CTG should be based on other tests of fetal well-being such as a biophysical profile or colour Doppler of the fetoplacental and uteroplacental circulation. It is probably of no practical value to do a CTG before 26 weeks of gestation. At such a preterm gestation, reasonable interventions cannot be offered in most units. The value of CTG at such gestations is only academic and as a documentation of life.

Antenatal Indications

The guiding principles for selecting women who need and may benefit from CTG are those whose fetuses are at risk for hypoxia. In the antenatal period, a number of such situations may be encountered. They are listed in Table 3.1. A more detailed discussion on common clinical situations and the use of CTG is presented below. The utility of CTG in reducing the risk of perinatal death has been assessed in a systematic Cochrane review.[3] The authors concluded that there is no evidence that antenatal CTG improves perinatal outcomes. However, there are

Table 3.1: Indications for CTG in the antenatal period			
Maternal systemic disease	**Maternal disease related to pregnancy**	**Clinical presentations**	**Fetal conditions**
Diabetes	Fetal death in previous pregnancy	Reduced fetal movements	Suspected Small for Gestational Age (SGA) fetus
Anemia	Hypertension, pre-eclampsia	False labour	Suspected or confirmed Fetal Growth Restriction (FGR)
Heart disease and essential hypertension	Gestational diabetes	Antepartum hemorrhage	Oligohydramnios
Autoimmune disease and vasculopathies	Postdates pregnancy	Prelabour rupture of membranes	Multiple pregnancy
Substance abuse			Fetal abnormalities
Obesity			Breech-before and after attempt of external cephalic version
Age over 40 years			False positive chromosomal serum screening tests in first or second trimester

methodological issues with the review. Only six studies were included with just over 2000 women studied in trials which were not of high quality. The review was underpowered to assess the outcome of perinatal death. As such, this remains a key question and further studies on the use of CTG, especially computerized CTG, in specific populations are warranted.

CTG in Women with Hypertension

Hypertension in pregnancy could be pre-existing or may arise as pre-eclampsia. The pathology of poor placentation, vaso-constriction and endothelial damage results in a fetoplacental unit that is prone to hypoxia. Antenatal CTG is used in women with hypertension with the thought that this is a high-risk population and the number of false positive results would be lower. This has not been borne out in clinical practice or in literature. In fact, in women with hypertension, the use of alphamethyldopa (a centrally acting antihypertensive agent) is likely to reduce FHR variability and produce a nonreactive CTG in a healthy fetus.[4] The

same is true if sedation or anxiolytics have been used. At present, most authorities recommend a weekly CTG for women with hypertension and twice weekly for women with severe disease or evidence of fetal compromise (growth restriction or oligohydramnios). These recommendations are opinion based. CTG can be used as an adjunct to other modalities for fetal well-being tests in this situation.

CTG in Women with Diabetes

Diabetes affects fetal well-being by changes in the metabolism or through vascular mechanisms. The metabolic change cause macrosomia, polyhydramnios and puts the fetus at risk for an otherwise unexplained fetal death. This is an unpredictable event. Vascular affection occurs in longstanding diabetes resulting in growth restriction and hypoxia. Fetal well-being tests are used to predict and prevent fetal deaths. This is usually initiated at 36 weeks gestation. The common practice is to advise CTG once or twice a week. In a study comparing CTG, biophysical profile and colour Doppler, CTG was inferior to umbilical artery colour Doppler in predicting fetal death and abnormal neonatal outcomes.[5] Therefore, it should not be the primary testing modality. It may be used as an adjunct tool.

CTG in Women with Reduced Fetal Movements

As pregnancy progresses, fetal movements have been looked upon as a sign of fetal health. A change in the pattern of fetal movements, especially a reduction or absence has been seen as a precedent of poor fetal outcomes, including fetal death.[6] The first step in a woman presenting with reduced fetal movements is to exclude fetal death. Having established that the fetus is alive, CTG is the first and almost universally performed fetal testing in a woman with reduced fetal movements. The initial test of 20 minutes may be extended to 80 minutes to achieve the criteria for a normal result. When the CTG is normal, it provides a great degree of reassurance. A recent Cochrane review found that the consensus opinion was to use CTG as a first line assessment tool for a woman

presenting with reduced fetal movements. They also opined that computerized CTG assessment and fetal arousal tests should be prioritized as areas for further research.[7] Figure 3.1 summarizes the clinical approach to a woman presenting with reduced fetal movements.

CTG in Women Presenting After the Due Date

The reducing efficiency of the placenta as the due date is passed is a cause for concern. It could lead to sudden, unexpected fetal deaths as well as a slower decline with hypoxia, oligohydramnios, abnormal umbilical Doppler velocimetries and eventually fetal death. There are ethnic differences and the timing of initiating fetal well-being tests should be individualized. Indian and Asian populations usually have an earlier placental maturation. It is prudent to initiate surveillance between 40 and 41 weeks of gestation. In the West, surveillance usually begins a week later. The most commonly used method of fetal well-being test is a biweekly CTG. This

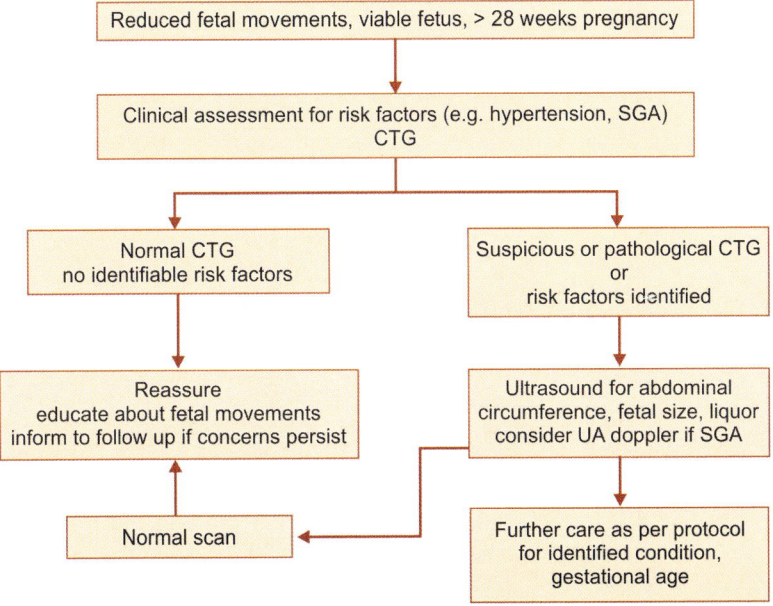

Fig. 3.1: Approach to a woman with the first episode of reduced fetal movements.

is endorsed by the NICE guidelines on routine antenatal care.[8]

CTG in Women with Suspected SGA Fetus, Oligohydramnios, Abnormal CD

In situations where there is a suspicion of fetal compromise with onset of hypoxia, the correct timing of delivery is critical. A delay in the delivery is valuable to allow steroids to be administered, maternal condition to be optimized and logistic preparations to be made. Certain ultrasound features such as absent or reversed end diastolic flow in the umbilical artery or very low amniotic fluid volume will require urgent delivery if the gestational age is viable. However, with less severe abnormalities such as high S/D ratio, moderate oligohydramnios, delivery may be delayed, especially if the period of gestation is less than 34 weeks. It is common to use CTG here, to provide instant information and usually reassurance about the fetal status, enabling the clinician to wait. However, most guidelines on the subject do not support the use of CTG as the only test in these situations since it is not associated with a better perinatal outcome. If CTG is used, interpretation and intervention should be based on short-term FHR variation from computerized analysis.[9]

CTG AS AN ADMISSION TEST

The Admission Test was a brief recording and documentation of the fetal heart trace when a woman presented in labour. The Admission Test was applied to women who had pregnancies thought to be at low risk for fetal hypoxia. The purpose was to identify fetuses that were already hypoxic at the time of admission in labour. The authors concluded that this simple, convenient test could detect fetal distress at the time of admission and unnecessary delay in intervention can be avoided. The test was also seen to have some predictive value of fetal well-being in the subsequent few hours of labour.[10] Since this pioneering work, much has changed in the

identification and assessment of women who are deemed to be at low-risk. A recent Cochrane review compared low-risk women who had an Admission Test CTG versus those who had intermittent auscultation.[11] It was found that women in the CTG group were more likely to have interventions such as continuous electronic fetal monitoring, fetal blood sampling, epidural analgesia and cesarean delivery. The risk of hypoxic ischemic encephalopathy and neonatal seizures was similar in both groups. The review was underpowered to detect a difference in perinatal mortality rates. It should be kept in mind that the studies included in the Cochrane review were conducted in well-equipped Western countries. The impact of the Admission Test may be different and more profound in a busy, understaffed labour ward in a developing country where fetal monitoring by intermittent auscultation many not be carried out as rigorously and a decision has to be made in terms of allocation of resources for fetal monitoring.

INTRAPARTUM INDICATIONS FOR CTG

A high-risk pregnancy may be identified before labour begins as outlined in Table 3.1. However, maternal systemic disease may be first identified at the time of admission in labour, especially hypertension, diabetes and anemia. The clinical examination and review are very important steps in the identification of these high-risk factors. In addition, new events and findings as labour progresses can change the status from low to high-risk. These situations are listed in Table 3.2. Even though, the evidence is not always clear, most guidelines recommend the use of CTG in high-risk situations.[12]

In women who are at low-risk, there is a consensus that CTG should not be offered as a routine. In a Cochrane review, CTG was compared to intermittent auscultation in low- risk women.[13] There was no difference

Table 3.2: Indications for continuous CTG monitoring that may arise in labour		
Maternal	**Labour process**	**Fetal**
Newly detected hypertension, proteinuria, diabetes, anemia	Prolonged rupture of membranes	Preterm or postdates fetus
Maternal tachycardia >120 bpm	Induction of labour	Abnormal admission CTG
Fever, sepsis or suspected chorioamnionitis	Delay in the first or second stage	Significant meconium stained liquor
Change in the nature of pain, indicating possibility of pain other than of uterine contractions	Oxytocin use for dysfunctional labour	
Fresh vaginal bleeding	Uterine hyperstimulation (hypertonus or tachysystole)	

in perinatal mortality rates, rates of cerebral palsy or cord blood analysis with acidosis. The use of CTG was associated with higher cesarean section and instrumental delivery rate. The only benefit of continuous CTG was a lower risk of neonatal seizures in this group as compared to those who were monitored with intermittent auscultation. At present, neonatal seizures are considered as a self-limiting condition which on follow-up at four years was not associated with any neurological deficit. Overall, it is thought to be of little significance, though it increases the need for neonatal intensive care.

CONCLUSIONS

Cardiotocography is a simple and convenient test of fetal well-being. Indications for its use are abundant since there are no known absolute contraindications. However, the use of this technology should be appropriate. The clinical scenario should be assessed as a whole and one factor should not govern the use of this technique. Selecting the fetus at risk for further assessment is the key to finding the correct indications for the use of cardiotocography and improving test performance.

REFERENCES

1. Manning FA. Antepartum fetal testing: a critical appraisal. Curr Opin Obstet Gynecol 2009;21: 348–52.

2. D'Antonio F, Bhide A. Antenatal Cardiotocography. In: Antenatal and Intrapartum Fetal Surveillance. (Eds): Arulkumaran S, Haththotuwa R, Tank JD, Tank PD. Universities Press, Hyderabad, 2013.

3. Grivell RM, Alfirevic Z, Gyte GML, Devane D. Antenatal cardiotocography for fetal assessment. Cochrane Database of Systematic Reviews 2015, Issue 9. Art. No.: CD007863. DOI: 10.1002/14651858.CD007863.pub4.

4. Rayburn WF, Motley ME, Zuspan FP. Conditions affecting nonstress test results. Obstet Gynecol 1982;59:490–3.

5. Bracero LA, Figueroa R, Byrne DW, et al. Comparison of umbilical Doppler velocimetry, nonstress testing, and biophysical profile in pregnancies complicated by diabetes. Journal of Ultrasound in Medicine1996;15:301–8.

6. Mangesi L, Hofmeyr GJ, Smith V. Fetal movement counting for assessment of fetal well-being. Cochrane Database of Systematic Reviews 2007, Issue 1. Art. No.: CD004909. DOI: 10.1002/14651858.CD004909.pub2.

7. Hofmeyr GJ, Novikova N. Management of reported decreased fetal movements for improving pregnancy outcomes. Cochrane Database Syst Rev. 2012 Apr 18;4:CD009148.

8. National Institute for Health and Care Excellence (NICE). Antenatal Care for uncomplicated pregnancies. Clinical Guideline. Published 2008, Revised 2017. Available at: nice.org.uk/guidance/cg62

9. Royal College of Obstetricians and Gynecologists (RCOG). The investigations and management of the Small for Gestational Age Fetus. Green Top

Guideline No.31. Published 2013, Revised 2014. Available at: https://www.rcog.org.uk/globalassets/documents/guidelines/gtg_31.pdf

10. Ingemarsson I, Arulkumaran S, Ingemarsson E, Tambyraja RL, Ratnam SS. Admission test: a screening test for fetal distress in labour. ObstetGynecol 1986; 68:800-6.

11. Devane D, Lalor JG, Daly S, McGuire W, Cuthbert A, Smith V. Cardiotocography versus intermittent auscultation of fetal heart on admission to labour ward for assessment of fetal well-being. Cochrane Database Syst Rev 2017; 1:CD005122. doi: 10.1002/14651858.CD005122.pub5.

12. Society of Obstetricians and Gynecologists of Canada (SOGC). Fetal Health Surveillance: Antepartum and Intrapartum Consensus Guidelines. J ObstetGynecol Canada 2007; 29:S4.

13. Alfirevic Z, Devane D, Gyte GM, Cuthbert A. Continuous cardiotocography (CTG) as a form of electronic fetal monitoring (EFM) for fetal assessment during labour. Cochrane Database Syst Rev. 2017 Feb 3;2:CD006066. doi: 10.1002/14651858.CD006066.pub3.

CHAPTER

4

Practical Tips for CTG Monitoring

Rubina Sohail, Bushra Haq, Madeeha Rashid

The purpose of CTG monitoring is to identify when there is concern about fetal well-being to allow interventions to be carried out before the fetus is harmed. The focus is on identifying FHR patterns associated with inadequate oxygen supply to the fetus. In general, the FHR patterns classified as normal are a reliable indication of fetal well-being. However, as far as the pathological patterns are concerned, up to 50% of FHR patterns classified as pathological, reflect physiological changes and can therefore be classified as false positives (false pathological). These false positive CTGs can lead antepartum to increased numbers of induced births and higher numbers of operative deliveries by cesarean section.

A normal FHR pattern is characterized by a baseline frequency between 110 and 150 beats per minute, presence of periodic accelerations, a normal heart rate variability with a bandwidth between 5 and 25 beats per minute and the absence of decelerations. In this regard, the accentuation of the 120 and 160 frequencies is misleading. It originates from the second half of the 19th century when the normal frequencies by von Winckel were stated to be between 120 and 160 beats per minute.[1]

The FHR pattern is abnormal when one or more of the following features are observed: a baseline frequency below 110 or above 150 beats per minute, absence of accelerations for more than 45 minutes, decreased or absent FHR variability and the existence of repeated variable or late decelerations. It is important to note that a baseline frequency between 100 and 110 can be considered as normal when the duration of pregnancy exceeds 41 weeks.[1]

OBJECTIVES

The objective of this chapter is to give some practical points, which will facilitate the use of CTG for the benefit of the patient, sort out the practical issues in performing and interpreting CTG and decrease judgement errors. It will also decrease unnecessary and untimely intervention of normal labour by cesarean section. For the purposes of convenience the practical points have been divided into various groups.

  i. Prerequisites of CTG
 ii. Components of CTG
iii. Interpretation of CTG
iv. Effect of medication on CTG
 v. CTG in special situations

I. Prerequisites of CTG

- When to do? Fetal well-being is assessed during antenatal period and labour.
  CTG machine—one must be familiar with CTG machine being used. Its type of paper used, CTG paper speed at 1 cm/minute, FHR range display set at 50–210 bpm. Health professionals should be aware that

machines from different manufacturers use different vertical axis scales, and this can change the perception of FHR variability.[2, 3]

- Position—sitting upright, lying on side, or at 45° angle all to avoid pressure on inferior vena cava.
- Correct placement of transducers—for FHR—on maternal abdomen best heard on anterior shoulder of fetus after assessing accurate position of the fetus and for uterine pressure—at the point of maximum uterine activity that is fundus.
- Basics:
  - Maternal pulse checked simultaneously
  - Initial 5 minutes must be observed to check the quality of CTG
  - Complete CTG means FHR and recording of uterine activity
  - If the trace is poor then change in position may improve it. Fetal movement recorded by the mother
  - Twin CTG: Simultaneous recording of two fetuses off set feature
  - Label CTGs with the mother's name, date, time commenced, hospital record number and include the maternal observations
  - Intrapartum events that may affect the FHR (e.g. starting or changing oxytocin, vaginal examination, obtaining fetal blood sample or insertion/siting an epidural) should be noted contemporaneously both on the CTG and in the maternal case notes, including date, time and signature.
- If normal criteria for CTG are met then further tracing may be stopped if patient has no pain, bleeding, etc.
- Monitor the CTG for abnormal trace if abnormality is found, extra vigilance is required.[4]

II. Components of CTG

Following are main features that should be systematically examined to assist with the interpretation of the CTG:

- Baseline rate
- Baseline variability
- Accelerations
- Decelerations
- Frequency and strength of contractions.

III. Interpretation of CTG

General Measures

- Having a series of CTG is better than one CTG alone as it can monitor the changing pattern of CTG. Timing of CTG can vary according to the need of individual patients.
- Do not make any decision about a woman's care in labour based on CTG findings alone, without correlating with the antenatal and intrapartum risk factors, the current well-being of the woman and unborn baby and the progress of labour.
- Ensure that the focus of care remains on the woman rather than the CTG trace.
- Remain with the woman to continue providing one-to-one support.
- Counsel the couple frequently about what is happening and take their preferences into account.

Specific Measures

Principles for intrapartum CTG trace interpretation:

- When reviewing the CTG trace, assess and document contractions and all 4 features of FHR: baseline rate; baseline variability; presence of accelerations, presence or absence of decelerations and concerning characteristics of variable decelerations, if present.
- If there is a stable baseline FHR between 110 and 160 beats/minute and normal variability, continue usual care as the risk of fetal acidosis is low.
- If it is difficult to categorize or interpret a CTG trace, obtain a review by a senior obstetrician.[5, 6]

- The presence of FHR accelerations, even with reduced baseline variability, is generally a sign that the baby is healthy.

FHR parameters and assessment criteria (Tables 4.1 to 4.3)

Intrapartum, the CTG reading must be constantly classified. The 30 minute segment with the highest number of suspicious or pathological FHR parameters must be analyzed and documented. If an assessment is classed as "suspicious", a repeat assessment should be done after 30 minutes and the number of suspicious parameters must be recorded. Here a number of conservative measures can be taken to clarify or improve the patterns (e.g. change of position, infusion).

If the reading is classified as "pathological", assessment must be continuous and recorded every 10 minutes including information on the number of suspicious parameters. In addition to various conservative measures

Table 4.1: FHR parameters and their definition (modified after ACOG, FIGO, SOGC, RCOG)[7,8]	
Term	**Definition**
Baseline (bpm)	Its mean FHR maintained over at least 10 minutes in the absence of accelerations or decelerations, given in beats per minute (bpm). For immature fetuses, mean FHR was in the upper range of variation. A progressive increase of FHR must be monitored carefully!
• Normal	Normal range: 110–160 bpm*
• Suspicious	Slight bradycardia: 100–109 bpm
	Slight tachycardia: 161–180 bpm without simultaneous accelerations
• Pathological	Severe bradycardia: <100 bpm
	Severe tachycardia: >180 bpm
Range (variability) (bpm)	Fluctuations in the fetal baseline rate occur 3–5 times per minute. The range is the difference in bpm between the highest and the lowest fluctuation during the most part of the 30 minute reading monitor strip.
• Normal	>5 bpm during the interval when no contractions occur
• Suspicious	<5 bpm and >40 minutes, but <90 minutes or >25 bpm

*Recent studies found that the physiological range for fetal heart rate at term was probably between 115 (4th percentile) and 160 beats per minute (96th percentile) (17, 105; EL II).
**<32nd week of gestation, rise of FHR >10 bpm or >½ range and >10 seconds. If accelerations are >10 minutes, this is considered a change in the baseline rate.

Table 4.2: Evaluation of individual FHR parameters (modified after ACOG, FIGO, SOGC, RCOG)[7,8]				
Parameter	**Baseline rate (bmp)**	**Range (bmp)**	**Decelerations**	**Accelerations**
Normal	110–160	≥5	None[1]	Present, sporadic[2]
Suspicious	100–109; 161–180	<5 ≥40 minutes >25	Early/variable decelerations, individual prolonged decelerations, up to 3 minutes	Present, periodical occurrence (with every contraction)
Pathological	<100 >180 sinusoidal[3]	<5 >90 minutes	Atypical variable decelerations, late isolated decelerations, prolonged decelerations >3 minutes	Absent >40 minutes (significance still unclear, evaluation questionable)

[1]FHR deceleration amplitude ≥15 bpm, duration ≥15 seconds [2]FHR acceleration amplitude ≥15 bpm, duration ≥15 seconds [3]sinusoidal FHR: ≥10 bpm, duration ≥10 minutes.

Table 4.3: FHR classification into normal, suspicious, pathological including need for action (based on FIGO)[9]

Category	Definition
Normal	All four assessment criteria are normal (no action required)
Suspicious	At least one assessment criterion is suspicious and all others are normal (need for action: conservative)
Pathological	At least one assessment criterion is pathological* or two or more are suspicious (need for action: conservative and invasive)

*does not apply to accelerations

Table 4.4: FHR is affected by the following factors[10]

Maternal	Fetoplacental	Fetal	Exogenous
Physical activity	Age of gestation	Movement	Noise
Posture-upright	Umbilical cord compression	Fetal behavioural states*	Medication
Uterine activity	Placental insufficiency	Stimulation to wake the fetus	Smoking
Body temperature (fever)	Chorioamnionitis	Hypoxemia	Drugs
Fluctuations in blood pressure			
• Vena cava syndrome			
• High diastolic BP reduces uterine perfusion			
Increased uterine tone	Placental abruption		

*From the 34th week of gestation on, fetuses show a cyclical change in heart rate patterns associated with changes in fetal behavioural states and fetal movement, shifting between resting (20–30 min) and activity (20–90 min); these changes are the surest sign of fetal well-being during the initial stage of labour and in the expulsive stage of labour. Mature fetuses spend around 80–90% of their time in one of these two defined states of activity. The remaining time they spend in a quiet state.

(e.g. tocolysis, attempts to wake the fetus, change of position, infusion, O_2 administration), Fetal Blood Analysis (FBA) should be done, if possible or useful. If no improvement in the CTG pattern can be achieved for one of the three important parameters or FBA shows pathological values, rapid delivery of the fetus is indicated. Factors influencing the CTG shown in Table 4.4.

IV. Effect of Medication on CTG

Medications can easily pass the placental barrier to reach the brain and other centers of circulatory regulation at high concentrations. They are mostly responsible for reducing the FHR variability (Table 4.5).

V. CTG in Special Situation

Continuous EFM in the Presence of Oxytocin

Intravenous oxytocin may be used to either induce or augment labour. Oxytocin to induce labour is often used in circumstances where the fetal risk is thought to be increased. Oxytocin is used to augment, inefficient uterine action, particularly in first-time mothers. If the FHR is normal, an oxytocin infusion may be increased until the woman is experiencing four to five contractions every 10 minutes. The oxytocin infusion rate can be titrated if contractions occur more frequently than five contractions in 10 minutes.[8]

Table 4.5: Effect of sedatives on FHR variability[11-13]	
Sedatives	
Anesthetic drugs—general and local	Reduces FHR variability
Antiepileptic drugs	Reduces FHR variability
Steroids—dexamethasone, betamethasone	Reduces FHR variability
Cocaine abuse	Reduces FHR variability
Magnesium sulphate	Reduces FHR variability
Beta mimetics—salbutamol, fenoterol	Increase in fetal heart rate and reduced variability Reversible after 5–7 days
Antihypertensive—beta blockers	Complete blockage of fetal sympathetic nervous system leading to decreased accelerations, pronounced brady-cardia and tachycardia Impair glucose mobilization.

If the FHR trace is classified as suspicious when an oxytocin infusion is in progress, a review by an experienced obstetrician should be requested. Once reviewed, the obstetrician may recommend that the oxytocin continues to be increased but only to a dose which achieves four to five contractions in 10 minutes.

Vaginal Birth After Cesarean (VBAC) Section

- Electronic Fetal Monitoring (EFM) should be offered, in established labour (>4 cm dilated with regular painful contractions)
- If using continuous EFM and there is a poor quality trace, then a Fetal Scalp Electrode (FSE) should be applied where possible
- Documentation is necessary.

Monitoring of Twins

Continuous external FHR monitoring of twin gestations during labour should preferably be performed with dual channel monitors that allow simultaneous monitoring of both FHRs, as duplicate monitoring of the same twin may occur and this can be picked up by observing almost identical tracings. Some monitors have embedded algorithms to alarm when this situation is suspected.

During the second stage of labour, external FHR monitoring of twins is particularly affected by signal loss, and for this reason some experts believe that the presenting twin should preferably be monitored internally for better signal quality if there are no contraindications to fetal electrode placement. Other experts believe that external monitoring of both twins is acceptable, provided that distinct and good quality FHR signals can be obtained.

Maternal Position and Oxygen Therapy

- During the presence of abnormal FHR patterns when a patient is lying supine she should be advised to adopt a left lateral position.
- Prolonged use of maternal reservoir facial oxygen therapy may be harmful to the baby and should be avoided. There is no research evidence evaluating the benefits or risks of short-term maternal reservoir facial oxygen therapy in suspected fetal compromise.[14]

Difficulty with FHR Detection

- In the event when there is difficulty in determining the fetal heart beat or an inability to detect a fetal heart with either a pinnard, handheld sonic aid or CTG machine; this should then be confirmed by ultrasound assessment.

- This must be confirmed by an obstetric registrar/consultant on call, with specific training in ultrasound or by a trained sonographer as soon as possible.
- If fetal death is suspected despite the presence of an apparently recorded FHR, offer real-time ultrasound assessment to check fetal viability.

Points to Remember

General

- Check that CTG machines are serviced regularly.
- Medical staff engaged in the use of electronic FHR monitoring should receive training and regular updating to maintain competency.
- Drills carried out to ensure appropriate response to different FHR patterns.

Specific

- Determine indication for fetal monitoring.
- Discuss fetal monitoring with the woman and obtain permission to commence.
- Perform abdominal examination to determine lie and presentation.
- Give the woman the opportunity to empty her bladder.
- The woman should be in an upright or lateral position (not supine).
- Check that accurate date and time has been set on the CTG machine.
- Maternal heart rate must be recorded on the CTG at commencement of the CTG to differentiate between maternal and FHR.
- Use standard definitions and objective assessment methods.
- Antepartum and on admission to the labour room (admission CTG) the usual (minimum) duration of recording is 30 minutes. Particularly in the third trimester of pregnancy the CTG should be obtained with the mother placed in a left lateral position to prevent vena cava syndrome.
- When using the FIGO score for the assessment of CTG readings, a reading of 30 minutes is necessary. The duration of the reading should be prolonged if the FHR pattern looks suspicious. A reduction of reading times to 10 minutes is possible with certain analysis methods (e.g. Dawes/Redman; Oxford system) if the results are confirmed. The maximum time for an Oxford CTG is 60 minutes.

Contd...

Contd...

- Monitoring frequency depends on the individual clinical risk confirmed by cardiotocography. It can range from a single reading done on an outpatient basis to several readings per day to continuous monitoring.
- If monitoring is done on an outpatient basis and monitoring sessions are more than four days apart, other monitoring systems with longer advance warning times (Doppler sonography, ultrasound evaluation of amniotic fluid volume, KCTG) should be additionally used.
- Registration of uterine contractions is done using an abdominal pressure transducer which converts the abdominal tension created by the contractions into a written signal, the tocogram, which provides information on both the frequency and the duration of contractions. If an external transducer is used, it will only show the relative strength of the contraction through a comparison of amplitudes, but, overall, this depiction of contraction strength is arbitrary.
- An external tocometer only indicates the presence of contraction. The amplitude of trace is not a measurement of intrauterine pressure. Decelerations cannot be timed with accuracy with extrauterine tocometer because correct placement is often difficult.
- External tocometer is placed at the most convex part of the uterus during contraction. The amplitude of contraction is less obvious in overweight mothers. Frequent alteration of recording site is essential to obtain a good display.
- Intra-amniotic pressure recordings are not necessary. It is generally recommended that the CTG should simultaneously record uterine contractions and FHR.
- Combining CTG with Doppler sonography in high-risk cohorts has reduced perinatal mortality by around 30%.
- If CTG findings are abnormal, fetal status should be additionally assessed with Doppler sonography, particularly in preterm infants.

REFERENCES

1. http://www.physicianspractice.com/articles/cardiotocography.
2. http://books.google.com.pk/books?ISBN...
3. Policy clinical guideline cardiotocography; SA Maternal and neonatal clinical Network 24 June 2015.

4. https://books.google.com.pk/books?id= S6YqDwAAQBAJ&printsec=frontcover&dq= assessing+fetal+well+being+a+practical+guide &hl=en&sa=X&ved=0ahUKEwiC0Py4vb PWAhVIXBQKHbl_ATMQ6wEIJjAA#v=one page&q=assessing%20fetal%20well%20being% 20a%20practical%20guide&f=false.

5. www.nice.org.uk/guidance/cg190/resources/ interpretation-of-cardiograph-traces.

6. Intrapartum care: NICE guideline CG190 (Feb. 2017) © National Institute for Health and Care Excellence 2017.

7. Royal College of Obstetricians and Gynae-cologists (RCOG). The Use of Electronic Fetal Monitoring, Evidence-based Clinical Guideline Number 8. RCOG Clinical Effectiveness Support Unit, London: RCOG Press; 2001. Available at URL: http://www.rcog.org.uk.

8. AyresdeCampos D, Spong CY, Chandraharan E. FIGO consensus guidelines on intrapartum fetal monitoring: Cardiotocography. International Journal of Gynecology & Obstetrics. 2015 Oct 1;131(1):13-24. Intrapartum care: NICE guideline CG190 (February 2017)©National Institute for Health and Care Excellence 2017.

9. https://www.ncbi.nlm.nih.gov/pubmed/ 26433401 Int J Gynaecol Obstet 2015;131(1):13–24. doi: 10.1016/j.ijgo.2015.06.020. FIGO consensus guidelines on intrapartum fetal monitoring: Cardiotocography.

10. S1-Guideline on the use of CTG during Pregnancy and lactation 2014 Aug 74;721–32.

11. Petrie RH, Yeh SY, Murata Y, et al. The effects of drugs on fetal heart rate variability. Am J Obstet Gynecol 1987;130:294–9 [PubMed].

12. Van Geijn HP, Jongsma HW, Doesburg WH, et al. The effect of diazepam administration during pregnancy or labor on the heart rate variability of the newborn infant. Eur J Obstet Gynecol Reprod Biol 1980;10:187–201 [PubMed].

13. Van Woerden EE, van Geijn HP. Amsterdam: Excerpta Medica; 1994. Factors influencing the fetal Heart Rate 211–20.

14. http://www.meht.nhs.uk/EasysiteWeb/ getresource.axd?AssetID=13650&type=full& servicetype=Attachment.

Uterine Activity Monitoring

Rubina Sohail, Asifa Noreen, Mariam Iqbal

Cardiotocography has two technical areas, one is the recording of the FHR in the form of a trace and the other is monitoring the uterine activity, which is also represented in the same trace. Both components of CTG are important and are complimentary to each other. It is recommended that both probes should be used for better interpretation of results during labour. Technically this was the basis of the stress test as well. CTG means recording the fetal heart beat and the uterine contractions during pregnancy. The probe used to measure the pressure is called the uterine transducer and machine used to perform the monitoring is called an electronic fetal monitor.

PARAMETERS OF UTERINE ACTIVITY

To measure and assess the uterine activity accurately, the following information is important:

- *Frequency:* The amount of time between the start of one contraction to the start of the next contraction.
- *Duration:* The amount of time from the start of a contraction to the end of the same contraction.
- *Intensity:* It is a measure of strength of the uterine contraction. With external monitoring, this necessitates the use of palpation to determine relative strength. With an Intrauterine Pressure Catheter (IUPC), this is determined by assessing actual pressures as graphed on the paper.

- *Interval:* The amount of time between the end of one contraction to the beginning of the next contraction.
- *Resting Tone:* This reflects the relaxation of the uterus is between contractions. With external monitoring, this necessitates the use of palpation to determine relative strength. With an IUPC, this is determined by assessing actual pressures as graphed on the paper.

UTERINE ACTIVITY MONITORING

One of the important components of monitoring of labour is monitoring the frequency, duration and intensity of uterine contractions. This becomes even more important in labour which is augmented by oxytocin.

Here it is important to understand what happens to the uterus during labour. The smooth muscle of the uterus contracts in a powerful manner with each contraction, building pressures up to 400 mm Hg (montvideo unit) at 10 cm cervical dilatation.[1] During a contraction, the maternal blood supply to the uteroplacental unit is compromised leading to transient hypoxia of fetus. In a healthy fetus, this state of transient hypoxia comes back to normal oxygenation as soon as the contraction ends. However, in certain conditions this may not be the case. This can happen frequently in a high-risk pregnancy but may be happening in low-risk pregnancies too. The most common cause in a normal pregnancy is prolonged labour and uterine contractions occurring too frequently (three contractions in ten minutes is considered adequate contractions). In a high-risk pregnancy where the fetus is already compromised, many a times, there is no reserve to tolerate transient hypoxia. This is reflected in changes in FHR in response to contractions.

Techniques for Monitoring Uterine Contraction

Traditionally uterine activity has been palpated and the duration, interval and intensity of uterine contractions documented as mild, medium or strong contractions (Fig. 5.1). To accurately assess and document uterine activity, two types of toco transducers can be used; the external and internal toco transducers.[2] An IUPC is a device placed into

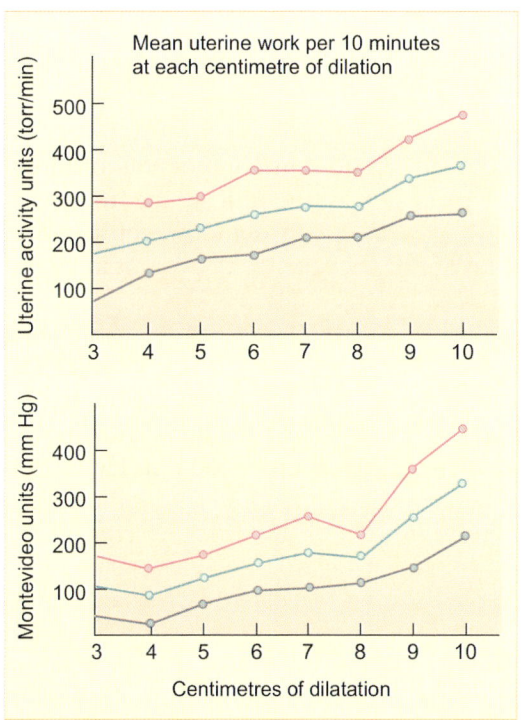

Fig. 5.1: Uterine work during labour measure in uterine activity units and Montevideo units. Both show increasing uterine work as labour proceeds.[1]

the amniotic space during labour in order to measure the strength of uterine contractions.

External Toco Transducers

The external pressure sensor or tocodynamometer is used for the monitoring of contraction. Here the tocodynamometer is strapped onto the abdomen over the fundus of uterus. This records the contractions, indirectly through sensing changes in skin tension arising from uterine muscular activity. Relative measure of contraction strength is done, calibrated in percentage over the range of 0–100%. This relative measurement gives an indication of the strength, duration and frequency of uterine contraction. The pressure recording in this method is not very accurate due to the limitation of the abdominal wall. The appearance of contractions by external monitoring may be affected not only by contraction strength but also by maternal habitus, position, gestational age, monitor location on the abdomen and tension of the abdominal belt, hence the amplitude of wave form on tocogram cannot be related to strength of uterine activity.[3]

Internal Intrauterine Pressure Monitoring Catheter

Intrauterine pressure catheters work by directly measuring pressure within the amniotic space using a pressure transducer at the tip of the catheter which allows for quantification of contraction strength. After connection to the appropriate cable, contractions are measured in mm Hg and displayed on the monitor in a graphic fashion (Fig. 5.2).

With an intrauterine pressure catheter in place, Montevideo Units (MVUs) can be calculated to assess for adequacy of labour in cases of suspected labour dystocia or during labour induction. MVUs are calculated by subtracting the baseline uterine pressure from the peak uterine pressure of each contraction in a 10-minute window of time and then taking the sum of these pressures. Two

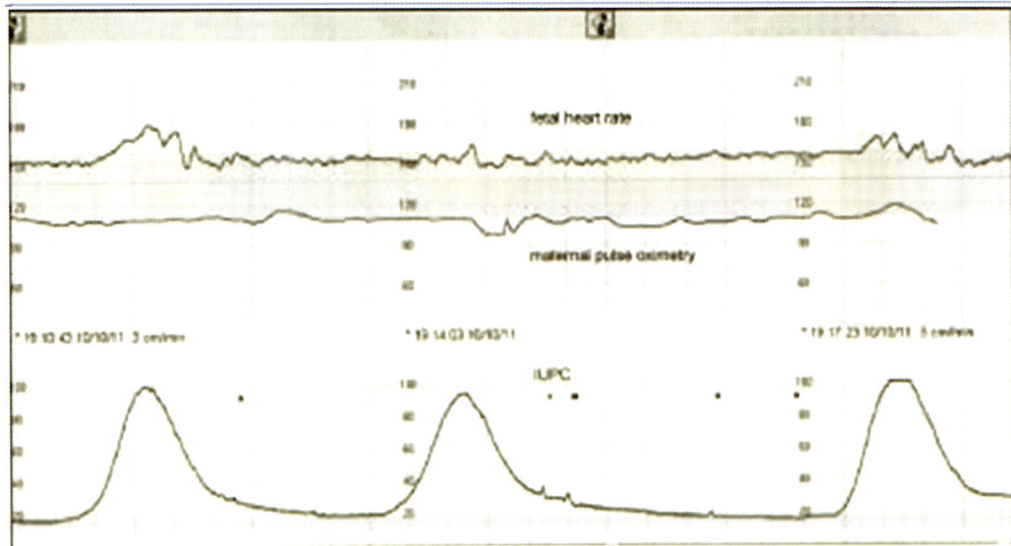

Fig. 5.2: intrauterine pressure catheter works by directly measuring the pressure within the amniotic space.[3]

hundred Montevideo units or more is considered adequate for normal labour progression.

An intrauterine pressure catheter is used, when quantification of contraction strength is desired, typically to assess the adequacy of spontaneous contractions in cases of arrested cervical dilation. It may also be used to facilitate titration of the oxytocin dosage during induction or augmentation of labour. An intrauterine pressure catheter can provide a more accurate assessment of contraction duration, length, and strength in patients in whom external tocodynamometry does not pick up contractions well, such as in obese patients. In cases of FHR decelerations, an IUPC can be used to clarify the relationship between the timing of the deceleration and the contraction.

Despite these advantages, routine use of intrauterine pressure catheters is not recommended. A large randomized trial of internal versus external tocodynamometry for monitoring labour showed no difference in rates of operative delivery or fetal outcomes between the two groups.

LIMITATIONS

Internal tocodynamometry is also costly, the sensors are of single use only and are expensive. Therefore, in the South Asian region, cost and availability are a serious limitation to use, its also an invasive procedure and its use is also associated with certain risks. These include introduction of infection, damage to maternal tissues and injury to the fetus. Therefore its use should be reserved for special circumstances.

Indications of Intrauterine Pressure Catheter (IUPC) Monitoring

- High-risk pregnancies
- Vaginal birth after cesarean section
- Failure to progress or prolonged labour
- Increased BMI (These are difficult to monitor with external toco transducer).

Contraindications of IUPC Monitoring

An intact fetal membrane is a contraindication to IUPC placement, as the desired location is within the amniotic space. Amniotomy just prior to IUPC placement is acceptable in the absence of contraindications to amniotomy.[4]

The Hemodynamics of a Uterine Contraction

- *Blood Volume:* An attempt has been made to estimate by indirect methods the amount of blood extruded from the uterus into the maternal venous reservoir during the early portion of the contraction cycle. The volume of blood may be in the range of 250 to 300 ml.
- *Blood Pressure:* The blood pressure rises quite consistently during a uterine contraction. Most commonly, the systolic pressure rises by 10 to 20 mm Hg, while the rise in diastolic pressure is also less.
- *Heart Rate:* During the early part of the contraction the heart rate tends to rise, followed by a substantial lowering of the rate by the time the contraction is at its maximum intensity.
- *Stroke Volume:* The stroke volume appears to drop slightly during the initial stages of the contraction cycle, after which it rises significantly above the base line level.
- The heart rate and stroke appear to maintain a somewhat reciprocal relationship throughout the contraction cycle.[5]

Changes in Uterine Blood Flow During Uterine Contractions

During spontaneous and oxytocin-induced labour, recordings were made of relative uterine blood flow (thermistor method), intrauterine pressure, femoral venous and arterial pressures, and maternal heart rate. In prelabour, uterine blood flow exhibits frequent irregular waves quite independent of the observed uterine activity. During labour, there is a characteristic relationship between uterine blood flow and uterine contractions. An initial decline in uterine blood flow precedes the contraction by about half a minute, followed by partial or complete recovery in the early contractile phase. A second drop begins as the intrauterine pressure reaches about 30 mm Hg, and the decline continues to the peak of the contraction or beyond. After the contraction is completed, the uterine blood flow recovers to its original level.[6]

Effect of Uterine Activity (UA) on Fetal Blood Flow

Uterine activity causes a decreased flow through the uterine artery. In the healthy uncompromised fetus, this will not cause fetal acidemia. The fetus has developed certain protection mechanisms to survive labour; (1) During a contraction, fetal preload increases and enables the fetus to maintain a constant blood flow through the umbilical artery and (2) UA increases the blood flow in the fetal middle cerebral artery, i.e. a brain sparing effect. The shortcoming of those protection mechanisms in the compromised fetus and in case of excessive UA increases the risk of adverse fetal outcome. The brain sparing effect will become more pronounced to compensate for the decreased umbilical artery blood flow and fetal oxygen saturation. Maintenance of normal UA, especially a sufficiently long relaxation time, is essential so that the supply of well oxygenated maternal blood to the intervillous space will be restored and the fetal cerebral oxygen saturation can remain stable.[7]

The NICHD nomenclature[9] defines uterine activity by quantifying the number of contractions present in a 10 minute window, averaged over 30 minutes. Uterine activity may be defined as:

- **Normal:** Less than or equal to 5 contractions in 10 minutes, averaged over a 30 minute window
- **Tachsystole:** More than 5 contractions in 10 minutes, averaged over a 30 minute window.[8]

Recent recommendations call for obstetricians to abandon the terms of "hyperstimulation" and "hypercontractility" in favour of

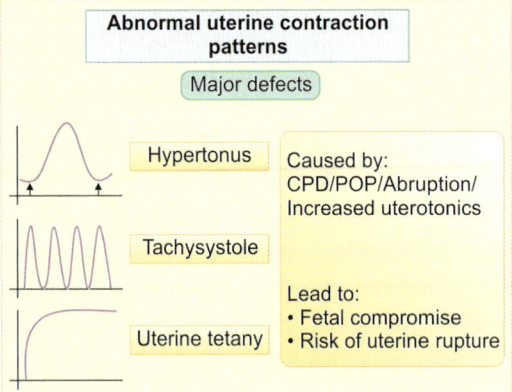

Fig. 5.3: Patterns of abnormal uterine activity[9,12]

the more rigidly defined term, "tachysystole" (TS).[9]

Abnormal Uterine Contraction Patterns

According to the level of tone, the abnormal uterine activity is classified into hypotonic and hypertonic uterine activity (Fig. 5.3).

Hypotonic Uterine Inertia

The uterine contractions are infrequent, weak and of short duration. In these women, labour is prolonged, uterine contractions are infrequent, weak and of short duration and there is poor cervical dilatation.

Hypertonic Uterine Inertia

Although definitions may vary among studies, most use the terms defined by the American College of Obstetricians and Gynecologists (1999a) to describe increased uterine activity as follows:

- **Uterine tachysystole** is defined as 6 contractions in a 10 minute period.
- **Uterine hypertonus** is described as a single contraction lasting longer than 2 minutes.
- **Uterine hyperstimulation** is when either condition leads to a nonreassuring FHR pattern.[10]

These findings are more common in primigravida and are characterized by prolonged labour. The uterine contractions are irregular and more painful. The pain is felt before and throughout the contractions with marked low backache often in occipitoposterior position. There is slow cervical dilatation.[11] High resting intrauterine pressure in between uterine contractions is detected by tocography (normal value is 5–10 mm Hg).

Points to Remember

- In monitoring of labour the three key areas are:
  a. Uterine contractions
  b. Fetal heart rate
  c. Dilatation of cervix and descent of presenting part
- For uterine contractions monitoring two methods are used:
  a. Maternal abdominal palpation is a crude method of assessing uterine contractions
  b. Tocodynometer is used to measure intensity, duration and interval of uterine contractions.
- Two type of toco transducers are used:
  a. **External transducer:** It has limited use as pressure is difficult to asses accurately. Used in low risk pregnancy
  b. **Internal transducer:** It is used in high risk pregnancy, VBAC, patient with increase BMI and failure to progress. It is an invasive procedure. It may cause introduction of infection and injury to fetus as amniotomy is a prerequisite for its use. Cost and availability are limitations for its use.

REFERENCES

1. Martin T. Whittle, Turn Bull's Obstetrics, The management and monitoring of labour. 2nd edition, Churchill living stone, 1997;571.
2. http://www.dremeilkamel.com.au/wp-content/uploads/2017/01/logo-reg.png
3. Robert Ogle/Jonathan Hyet, High risk Pregnancy, Screening for spontaneous preterm labour and delivery. 3rd edition, Elsevier 2006;1249.
4. Nan G O'Connell (Author), Carl V Smith (Chief Editor). Intrauterine Pressure Catheter Placement.
5. Charles H Hendricks, Cleveland Ohio. The hemodynamics of a uterine contraction. American Gynecological Society Transactions of the Eighty-first Annual Meeting.

6. Vladimir Brotanek, Hendricks, Yoshida. Changes in uterine blood flow during uterine contractions. American Journal of Obstetrics and Gynecology.

7. Bakker PC, van Geijn HP. Uterine activity: implications for the condition of the fetus. J Perinat Med 2008;36(1):30–7.

8. Macones George A, Hankins Gary DV, Spong Catherine Y. Hauth John Moore, Thomas . "The 2008 National Institute of Child Health and Human Development Workshop Report on Electronic Fetal Monitoring". Obstetrics & Gynecology. 112(3):661–6. PMID 18757666. doi:10.1097/AOG.0b013e3181841395.

9. Heuser CC[1], Knight S, Esplin MS, Eller AG, Holmgren CM, Manuck TA, Richards D, Henry E, Jackson GM. Tachysystole in term labor: incidence, risk factors, outcomes, and effect on fetal heart tracings. Am J Obstet Gynecol. 2013 Jul;209(1):32.e1-6. doi: 10.1016/j.ajog.2013.04.004. Epub 2013 Apr 6.

10. http://www.gfmer.ch/Obstetrics_simplified/abnormal_uterine_action.htmObstetrics Simplified—Diaa M. EI-Mowafi. Abnormal Uterine Action.

11. https://en.wikipedia.org/wiki/Uterine_tachysystole

12. https://www.slideshare.net/drmdsadiq/case-capsules.

CHAPTER

6

Normal CTG Interpretation

Asanka Jayawardane, Rohana Haththotuwa

Cardiotocography (CTG) is an electronic method of capturing and recording the fetal heart rate (FHR) and uterine activity (contractions) to evaluate fetal well-being during pregnancy and labour. When performed antepartum it is termed Non Stress Test (NST), i.e. the fetus is being tested in the absence of uterine activity, a 'stress-free' situation as opposed to intrapartum monitoring with uterine contractions.

RECORDING CTG

CTG is most commonly carried out externally. Two external transducers are used simultaneously to monitor the FHR and uterine contractions. The external tocographic pressure transducer is placed halfway between the uterine fundus and the umbilicus where the change of abdominal girth anteroposteriorly is maximal with uterine contractions. The ultrasound transducer to measure the FHR is placed on the abdomen where the fetal heart could be maximally heard which is usually directly over the chest or anterior shoulder of the fetus.

Technology has evolved to capture the fetal electrocardiogram via electrode placed on fetal scalp or via the maternal abdomen for FHR monitoring. Measurement of the intrauterine pressure has been made possible via a strain-gauge or fluid filled intrauterine catheters.

STANDARDIZATION OF RECORDING

There are no internationally agreed guidelines. CTG paper speed at 1 cm/minute is recommended by the UK. Some countries (USA) use paper speeds of 3 cm/min. Most machines can record at both speeds as preferred by the person recording. It is recommended that sensitivity displays at 20 beats per minute/cm and set FHR range display at 50–210 bpm. Using different vertical axis scales or paper speeds may go unnoticed at interpretation and this can change the perception of fetal well-being.

Ensure date and time are correct on commencement of CTG and label all CTGs with the mother's name, date, time commenced, hospital record number and include the maternal observations (HR/temperature). Intrapartum events that may affect the FHR (e.g. starting or changing oxytocin regimen, vaginal examination, obtaining fetal blood sample or insertion/siting an epidural) should be noted contemporaneously both on the CTG and in the maternal case notes, including date, time and signature of the attending doctor or midwife (RANZCOG 2014).[1]

THE FEATURES OF A CTG

Baseline FHR

Baseline FHR is the mean rate when this is stable, excluding accelerations and decelerations over a ten minute period and is expressed as beats per minute (bpm). The FHR of preterm fetuses may be in the upper range due to later maturation of the parasympathetic system. A FHR within the normal baseline with accelerations and normal baseline variability (and without decelerations) is not associated with hypoxia. The normal baseline rate is 110–160 beats per minute.

Baseline Variability

The baseline variability is the minor baseline FHR fluctuations measured by gauging the difference in bpm between the highest peak and lowest trough of fluctuation in one minute segments of the trace between contractions. **Variability occurs as a result of the interaction between the sympathetic and parasympathetic nervous system.** The normal baseline variability is between 5 and 25 bpm. Variability is said to be reduced when it is <5 bpm and may be associated with fetal hypoxia with markedly reduced baseline variability especially if associated with late decelerations. Caution should be exercised in interpreting variability in the presence of an external transducer. Reduced variability of less than 5 bpm may be normal in a term fetus if seen as part of sleep phase of 'cycling' (up to 20 to 40 minutes duration and occasional slight rise in baseline). It may also occur secondary to drugs such as pethidine. In some literature a baseline variability of less than 3 is termed absent variability and may denote a suppressed nervous system due to hypoxia, medication or infection (Sweha et al. 1999).[3]

Saltatory FHR variability is defined as fetal heart amplitude changes of more than 25 beats per minute. This may occur in a rapidly evolving hypoxia if such variability persists over a ten minute period and/or with repetitive decelerations. Such patterns are recognized in the second stage of labour (RCOG 2001).[6]

Accelerations

Accelerations are abrupt transient increases in FHR of 15 bpm or more above the baseline and lasting for >15 seconds. They are ususally associated with fetal movements or somatic nervous system activity. It is strongly related to normal fetal oxygenation and non-acidotic fetus. Accelerations in the preterm fetus may be of lesser amplitude (>10 bpm) and shorter duration. The significance of no accelerations on an otherwise normal CTG is unclear.

Decelerations

Decelerations are transient episodes of decrease of FHR below the baseline rate of more than 15 bpm lasting at least for >15 seconds. Decelerations may be further classified as early, variable, complicated variable, late and prolonged decelerations.

Overall Impression-classification of CTG

NICE (UK), guidelines recommend a systematic classification of a CTG for in labour fetal monitoring as described in Tables 6.1 and 6.2 (NICE 2017).[2]

How to reat a CTG: A systematic approach
To interpret a CTG a structured method of assessing its various characteristics must be employed. This tends to reduce the inter and intra personal variability in interpretation of CTG. A simple pattern recognition and unscientific classification may lead to costly errors of judgement.

The most popular structure can be remembered using the acronym **DR C BRAVADO**

DR – **D**efine **R**isk

C – **C**ontractions

BRa – **B**aseline **R**ate

V – **V**ariability

A – **A**ccelerations

D – **D**ecelerations

O – **O**verall impression

Table 6.1: NICE guidelines for interpretation of intrapartum CTGs (NICE 2017)[2]			
Description	**Feature**		
	Baseline (beats/minute)	**Baseline variability (beats/minute)**	**Decelerations**
Reassuring	110 to 160	5 to 25	None or early Variable decelerations with no concerning characteristics* for less than 90 minutes
Non-reassuring	100 to 109[†] or 161 to 180	Less than 5 for 30 to 50 minutes or More than 25 for 15 to 25 minutes	Variable decelerations with no concerning characteristics* for 90 minutes or more or Variable decelerations with any concerning characteristics* in up to 50% of contractions for 30 minutes or more or Variable decelerations with any concerning characteristics* in over 50% of contractions for less than 30 minutes or Late decelerations in over 50% of contractions for less than 30 minutes, with no maternal or fetal clinical risk factors such as vaginal bleeding or significant meconium
Abnormal	Below 100 or above 180	Less than 5 for more than 50 minutes or More than 25 for more than 25 minutes or Sinusoidal	Variable decelerations with any concerning characteristics* in over 50% of contractions for 30 minutes (or less if any maternal or fetal clinical risk factors (*see* above) or Late decelerations for 30 minutes (or less if any maternal or fetal clinical risk factors) or Acute bradycardia, or a single prolonged deceleration lasting 3 minutes or more

*Regard the following as concerning characteristics of variable decelerations: lasting more than 60 seconds; reduced baseline variability within the deceleration; failure to return to baseline; biphasic (W) shape; no shouldering.
[†]Although a baseline fetal heart rate between 100 and 109 beats/minute is a non-reassuring feature, continue usual care if there is normal baseline variability and no variable or late decelerations.

The Normal CTG (Fig. 6.1)

The normal CTG is associated with a low probability of fetal compromise and has the following features:

- Baseline rate 110–160 bpm
- Baseline variability of 6–25 bpm
- Accelerations of 15 bpm for 15 seconds
- No decelerations

All other CTGs are by this definition are abnormal and require further evaluation along with the full clinical picture (Morris 2006).[7]

The following features are unlikely to be associated with fetal compromise when occurring in isolation:

- Baseline rate 100–109 bpm
- Absence of accelerations
- Early decelerations in the late first and second stages of labour
- Variable decelerations without complicating features.

Category	Definition	Management
Table 6.2: Management based on interpretation of CTG (NICE 2017)[2]		
Normal	All features are reassuring	• Continue CTG (unless it was started because of concerns arising from intermittent auscultation and there are no ongoing risk factors; see recommendation 1.10.8) and usual care • Talk to the woman and her birth companion(s) about what is happening
Suspicious	1 non-reassuring feature and 2 reassuring features	• Correct any underlying causes, such as hypotension or uterine hyperstimulation • Perform a full set of maternal observations • Start 1 or more conservative measures* • Inform an obsterician or a senior midwife • Document a plan for reviewing the whole clinical picture and the CTG findings • Talk to the woman and her birth companion(s) about what is happening and take her preferences into account
Pathological	1 abnormal feature or 2 non-reassuring features	• Obtain a review by an obstetrician and a senior midwife • Exclude acute events (for example, cord porlapse, suspected placental abruption or suspected uterine rupture) • Correct any underlying causes, such as hypotension or uterine hyperstimulation • Start 1 or more conservative measures* • Talk to the woman and her birth companion(s) about what is happening and take her perferences into account • If the cardiotocograph trace is still pathological after implementing conservative measure: ♦ obtain a further review by an obstetrician and a senior midwife ♦ offer digital fetal scalp stimulation (see recommendation 1.10.38) and document the outcome • If the cardiotocograph trace is still pathological after fetal scalp stimulation: ♦ consider fetal blood sampling ♦ consider expending the birth ♦ take the woman's perferences into account

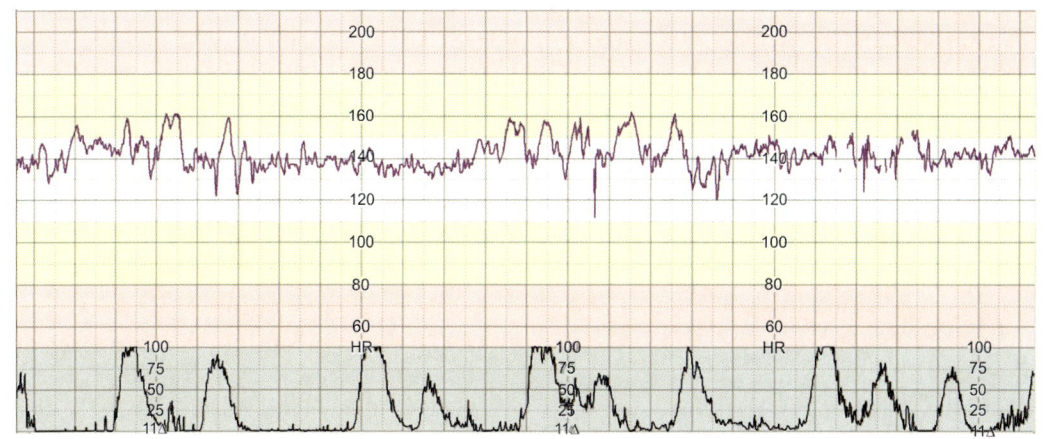

Fig. 6.1: Normal CTG with a baseline rate of 140 bpm, baseline variability of 5–25 bpm, several accelerations and no decelerations.

The following features may be associated with significant fetal compromise and require further action (Parer JT et al. 2006):[5]

- Baseline fetal tachycardia >160 bpm
- Reduced or reducing baseline variability (3–5 bpm)
- Rising baseline fetal heart rate
- Complicated variable decelerations—the depth and breadth increases with reduction of inter decelration intervals with progressing hypoxia
- Late decelerations especially with markedly reduced baseline variability
- Prolonged decelerations >3 minutes.

Two non reassuring features makes a CTG to be classified as abnormal (Fig. 6.2): The following features are likely to be associated with significant fetal compromise and require immediate management, which may include urgent delivery (RCOG 2001):[6]

- Prolonged bradycardia (>6 minutes)
- Absent baseline variability (<3 bpm)
- Sinusoidal pattern
- Complicated variable decelerations with reduced or absent baseline variability
- Late decelerations with reduced or absent baseline variability Fig. 6.3.

PHYSIOLOGY OF FHR MONITORING

The fetal heart is detectable by transvaginal ultrasound as early as 4 weeks after conception. At this stage the mean FHR is about 100 bpm. Thereafter mean FHR progressively rises, reaching 140–150 by 10 weeks menstrual age (8 weeks post conception), and levels off at that rate by the start of the second trimester.

From 14 weeks to term there is a progressive fall in the mean baseline FHR which is unaffected by whether the fetus is Active or Quiescent.

This lowering of the baseline rate with gestation is a reflection of the fact that the sympathetic autonomic nervous system matures earlier than the parasympathetic system.

The baseline variability of the FHR in early pregnancy is low and increases with gestation and changes are seen with the behavioural state of the fetus. Thus, over the second half of pregnancy the baseline variability increases progressively during fetal activity; this increase is less marked during fetal quiescence and declines from 30 weeks onwards.

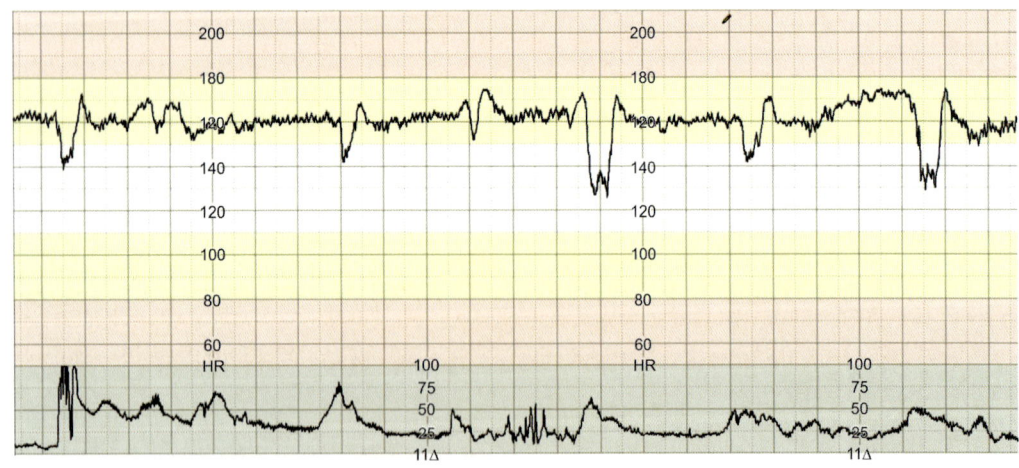

Fig. 6.2: Simple variable decelerations with baseline tachycardia of 160 bpm makes this trace to be classified as pathological. The presence of normal baseline variability and absence of atypical variable decelerations allows an opportunity to observe.

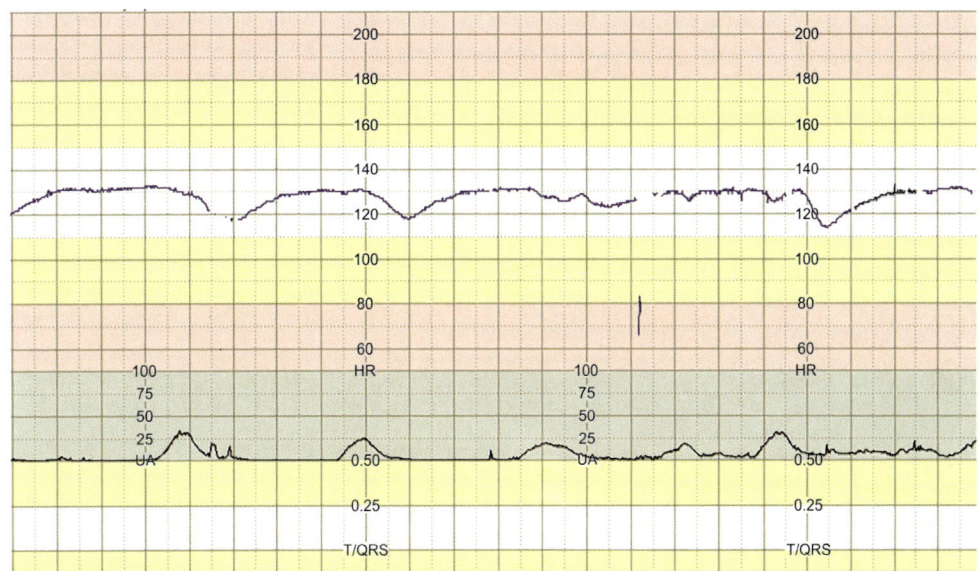

Fig. 6.3: Although baseline rate is in the normal range, the absence of baseline variability and late decelerations makes it a pathological trace needing immediate intervention.

At 30 weeks there is not much difference in baseline variability between fetal activity and quiescence. However, after >35 weeks when the fetus is quiet the baseline variability is reduced to <5 bpm and is 5–25 bpm with fetal activity.

Accelerations develop and increase with advancing gestation and they occur with fetal movements.

Over the last trimester the normal fetus commonly manifests three behavioural states defined on the basis of three parameters – the FHR, eye movements and body/limb movements. The nomenclature is derived from the five behavioural states manifest by newborns.

1F: Low variability baseline FHR, no eye movements, occasional 'startle' body movement with brief rise in FHR, present on average for ~30% of the time, maximum duration <40 min (can be longer but this is a reasonable definition for use in practice).

2F: High variability baseline FHR, eye movements present, many fetal movements with FHR accelerations, present on average for ~60% of the time, maximum duration up-to 90 mins.

4F: Sustained accelerations with occasional return to baseline FHR, rapid eye movements, continuous fetal movements, present on average for ~10% of the time, maximum duration up to 120 mins.

In the CTG recording, the sleep activity periods are defined by segments of normal baseline variability with accelerations alternating with segments of reduced baseline variability and the absence of accelerations and is termed 'cycling' (Fig. 6.4.).

COMPUTER AIDED FHR MONITORING

Fetal monitoring with cardiotocogram has poor intra and inter observer agreement. Classifications, scoring systems and guidelines introduced to increase consistency have not improved this problem. Overall agreement in classification of CTG tracings by "experts" to normal, suspicious and pathological categories according to guidelines was fairly poor. The consistency was even less for intrapartum recordings.

Computerised FHR recording analysis was developed to aid reliable assessment of the CTG (Fig. 6.5). The system 8000 (current

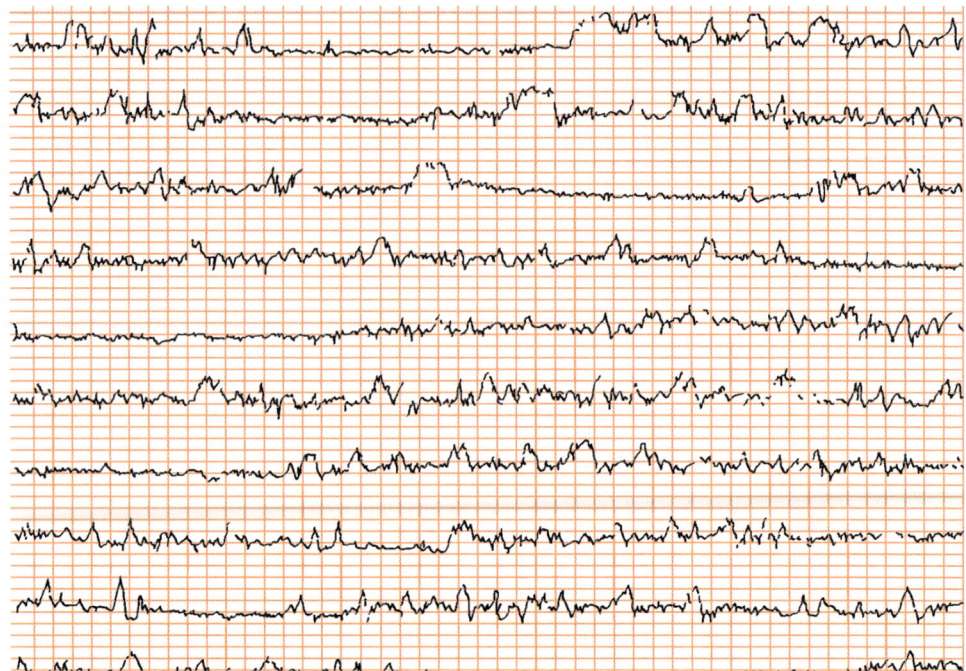

Fig. 6.4: The fetal heart rate pattern from the beginning to end of labour showing the 'active' and 'quiet' epochs, i.e. 'cycling' indicative of normal behavioural state.

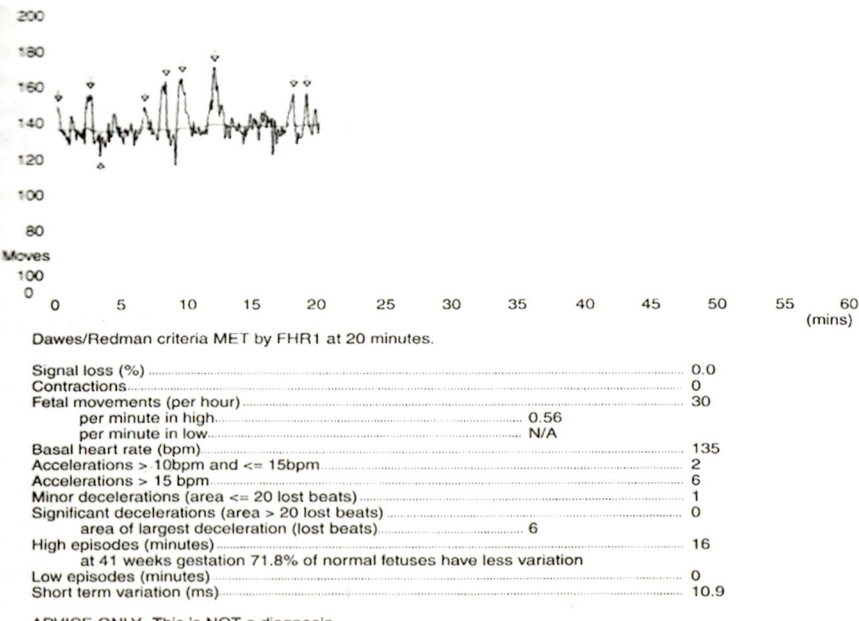

Dawes/Redman criteria MET by FHR1 at 20 minutes.

Signal loss (%)	0.0
Contractions	0
Fetal movements (per hour)	30
per minute in high	0.56
per minute in low	N/A
Basal heart rate (bpm)	135
Accelerations > 10bpm and <= 15bpm	2
Accelerations > 15 bpm	6
Minor decelerations (area <= 20 lost beats)	1
Significant decelerations (area > 20 lost beats)	0
area of largest deceleration (lost beats)	6
High episodes (minutes)	16
at 41 weeks gestation 71.8% of normal fetuses have less variation	
Low episodes (minutes)	0
Short term variation (ms)	10.9

ADVICE ONLY. This is NOT a diagnosis.

Fig. 6.5: Computerised AN CTG –Dawes Redman Criteria for a normal CTG. Should have one episode of high variation. STV > 3.0 ms. if <4.5 ms the averaged LTV > 3rd centile for gestational age. No high frequency sinusoidal rhythm. At least one FM or 3 accelerations. No decelerations > 20 lost beats if recording is <30 mins and 21–100 beat loss if recording is > 30 mins/ No decelerations >100 beat loss. BLR 116–160 bpm if recording is <30 mins.

Sonicaid ("FetalCare") is one of the pioneer systems with a large database backup. The Fetal Care system analyses the basal heart rate, accelerations, decelerations, Long-Term Variation (LTV) and Short-Term Variation (STV) and also confirms that there is no sinusoidal pattern present indicative of severe fetal anemia. A parameter that cannot be assessed visually but can be calculated in computerized systems is the STV. It is calculated as the difference of the average pulse intervals in two consecutive 3.75 second time epochs. The analysis is compared with Dawes Redman criteria and the result is given as criteria met or not met. Irrespective of the time taken to meet the criteria the fetus is considered well if they are met. A maximum of 60 minutes is allowed to distinguish from a quiet sleep cycle of the fetus.

The advantages of the system include its ability to overcome observer variability and being independent of the degree of experience of the staff. It also considers the STV which cannot be visually assessed. The method itself is non-invasive and highly acceptable to the patients.

EVIDENCE-BASED FETAL MONITORING

Antenatal fetal monitoring, though employed widely in practice, has not been shown to be an effective stand alone tool for detecting or monitoring an at risk fetus (Grivell et al 2012, Grivell et al 2015).[4,8]

Cochrane review (Alfirevic et al 2017)[9] on continuous electronic fetal monitoring in labour revealed that CTG was associated with fewer neonatal seizures but it is unclear if it had any impact on long-term neurodevelopmental outcomes. No clear differences in cerebral palsy, infant mortality or other standard measures of neonatal well-being could be shown. There was also no advantages on long-term outcomes.

On the other hand many investigators have demonstrated that fetal monitoring with continuous CTG was associated with the higher rates of medical interventions including cesarean sections and instrumental vaginal births.

Therefore NICE 2017[2] recommends not to offer cardiotocography to women at low-risk of complications (including women who have non-significant meconium if there are no other risk factors) in established labour and recommends use of intermittent auscultation of the FHR to women at low-risk of complications in established first stage of labour.

REFERENCES

1. The Royal Australian and New Zealand College of Obstetricians and Gynaecologists (RANZCOG): Intrapartum Fetal Surveillance. Clinical Guidelines – Third Edition; 2014 (Level IV). Available at URL: *http://www.ranzcog.edu.au.*

2. NICE guidelines http://www.nice.org.uk/nice-media/live/11837/36273/36273.pdf Intrapartum care for healthy women and babies Clinical guideline [CG190] Published date: December 2014 Last updated: February 2017.*https://www.nice.org.uk/guidance/cg190/evidence*

3. Sweha A, Hacker TW, Nuovo J. Interpretation of the electronic fetal heart rate during labor.Am Fam Physician. 1999;59(9):2487–500. http://www.aafp.org/afp/990501ap/2487.html

4. Grivell RM, Alfirevic Z, Gyte GML, Devane D. Antenatal cardiotocography for fetal assessment. Cochrane Database of Systematic Reviews 2012, Issue 12. Art. No.: CD007863. DOI: 10.1002/14651858.CD007863.pub3. (Level I). Available from URL: http://www.mrw.interscience.wiley.com/cochrane/clsysrev/articles/CD007863/pdf_st andard_fs.html.

5. Parer JT, King T, Flanders S, Fox M, Kilpatrick SJ. Fetal academia and electronic fetal heart rate patterns: Is there evidence of an association? J Maternal-Fetal and Neonatal Med 2006;19: 289–94.

6. Royal College of Obstetricians and Gynaecologists (RCOG). The Use of Electronic Fetal Monitoring, Evidence-based Clinical Guideline Number 8. RCOG Clinical Effectiveness Support Unit, London: RCOG Press; 2001. Available at URL: http://www.rcog.org.uk

7. Morris D. EFM Master Tutor. A practical approach to the interpretation of CTG; 2006. Version 6.0. Available at URL: http://www. response-education.com.au/

8. Grivell RM, Alfirevic Z, Gyte Gillian ML, Devane D, Grivell RM. "Antenatal cardiotocography for fetal assessment". The Cochrane Database of Systematic Reviews 2005;(9): CD007863. PMID 26363287. doi:10.1002/14651858.CD007863.pub4

9. Alfirevic Z, Devane D, Gyte Gillian MI, Cuthbert A. "Continuous cardiotocography (CTG) as a form of electronic fetal monitoring (EFM) for fetal assessment during labour". The Cochrane Database of Systematic Reviews (3 February 2017) 2: CD006066. ISSN 1469–493X. PMID 28157275. doi:10.1002/ 14651858. CD006066.pub3.

Antenatal Monitoring

Ameya C Purandare, Madhuri Mehendale

Antenatal monitoring comprises of various techniques to ensure fetal well-being and usually focuses on fetal biophysical findings that include fetal heart rate, fetal body movements, fetal breathing and amniotic fluid production. These help us achieve the goals recommended by the American College of Obstetricians and Gynaecologists and the American Academy of Paediatrics (2012),[1] which include prevention of fetal death and avoidance of unnecessary interventions.

An ideal antenatal fetal monitoring test aims to reduce the fetal and neonatal outcomes of asphyxia listed in Table 7.1.[2]

Antenatal fetal monitoring techniques mentioned below may be used simultaneously or in a hierarchical fashion depending on risk factors and resource settings.

1. Fetal movement counting
2. Non-stress test
3. Contraction stress test
4. Biophysical profile and/or amniotic fluid volume
5. Maternal uterine artery Doppler
6. Fetal umbilical artery Doppler

INDICATIONS FOR ANTEPARTUM FETAL SURVEILLANCE

Perinatal morbidity and/or mortality due to fetal asphyxia have been shown to be increased among women with conditions identified below.[2]

Maternal History in Previous Pregnancies

- Hypertensive disorder of pregnancy
- Placental abruption
- Fetal intrauterine growth restriction
- Stillbirth.

Current Pregnancy

- Maternal Post-term pregnancy (more than 294 days or more than 42 weeks)
- Hypertensive disorders of pregnancy
- Pre-pregnancy diabetes
- Insulin requiring gestational diabetes
- Preterm premature rupture of membranes
- Chronic (stable) abruption
- Isoimmunization
- Abnormal maternal serum screening in absence of confirmed fetal anomaly
- Vaginal bleeding
- Morbid obesity.

Table 7.1: Adverse fetal and neonatal outcomes associated with antepartum asphyxia (Asphyxia is defined as hypoxia with metabolic acidosis)	
Fetal outcomes	**Neonatal outcomes**
Stillbirth	Mortality
Metabolic acidosis at birth	Metabolic acidosis
	Hypoxic renal damage
	Necrotizing enterocolitis
	Intracranial hemorrhage
	Seizures
	Cerebral palsy
	Neonatal encephalopathy

Fetal

- Advanced maternal age
- Pregnancy following the use of assisted reproductive technologies
- Decreased fetal movement
- Intrauterine growth restriction
- Suspected oligohydramnios/polyhydramnios
- Multiple pregnancy
- Preterm labour.

Methods of Antenatal Monitoring

Methods used would largely depend on type of settings:

Low Resource settings (Primary Health Care Centres)

1. Symphysiofundal height[3]

This refers to the distance measured in centimetres on the longitudinal axis of the abdomen from the top of the fundus to the upper border of the symphysis pubis. A measurement discrepancy of more than 2 cm can be suggestive of a fetus that is small/large for gestational age, multiple pregnancy, or an inaccurate estimated due date. Other causes include molar pregnancies, polyhydramnios/oligohydramnios, an oblique or transverse lie. Between 20 and 34 weeks of gestation the height of the uterus correlates closely with measurements in centimetres, however obesity has been shown to distort the accuracy of these measurements. Also towards the end of pregnancy measurements become less accurate due to the descent of the fetal presenting part into the maternal pelvis.

> **Key Points**
> - Symphysiofundal height it to be performed at every scheduled antenatal visit after 24 weeks of gestation.
> - Symphysiofundal height must always be documented in centimetres.
> - If there is a discrepancy in size and gestation of >2 cm, further surveillance is needed.
> - Using fundal height alone, fetal growth restriction may be undiagnosed in up to a third of cases (ACOG, 2013)[4]

2. Fetal Movement Counting

Decreased placental perfusion and fetal acidemia and acidosis are associated with decreased fetal movements. This is the basis for maternal monitoring of fetal movements or "the fetal kick count test." The concept of counting fetal movements is cost effective and easily feasible, since it requires no technology and is available to all women, even in low resource settings.

Review of the Evidence

1. In a review of the literature since 1970 on fetal movement counting in several countries, frozen analyzed 24 studies and performed several meta-analyses on the data. His major findings included the following.[5]
   - In high-risk pregnancies, the risk for adverse outcomes in women with decreased fetal movements increased mortality, chances of IUGR, APGAR <7 at 5 minutes and need for emergency delivery.
   - There was a trend to lower fetal mortality in low-risk women in the fetal movement groups versus controls, although this difference was not statistically significant.
2. Grant et al. (1989) performed a study on 68000 pregnancies between 28 and 32 weeks.[6] They studied maternally perceived fetal movements and pregnancy outcome. They concluded that informal maternal perceptions were as valuable as formally recorded fetal movement.

The only antenatal surveillance technique recommended for all pregnant women, with and without above risk factors, is maternal awareness of fetal movements.

Methods of Fetal Movement Count

A variety of methods have been described, which are usually variations on the methodologies of two early studies.
- The Cardiff method, first reported by Pearson and Weaver[7] suggests a count to

10 movements in a fixed time frame. The original study required counting for 12 hours. Modified protocols include those of Liston (count to 6 hours) and Moore[8] (count to 2 hours).

- The Sadovsky[9] method suggests a count of movements in a specific time frame (usually 30 minutes to two hours).

There are no studies comparing the effect on outcome of using different fetal movement count charts. Ideally, the testing should be performed for the shortest time possible to identify fetuses at risk. A short observation period allows women to concentrate on the fetal movement count while minimizing any imposition on routine daily activity. Figure 7.1 depicts the use of the Fetal Movement Count in fetal surveillance.

High Resource Settings (Secondary and Tertiary Level Health Centres)

Antenatal surveillance can be done by:
- Non-stress test
- Contraction stress test
- Biophysical profile and/or amniotic fluid volume
- Maternal uterine artery Doppler, and fetal umbilical artery Doppler.

1. Nonstress Test (NST)

The NST is based on the principle that the fetal heart will accelerate with fetal movement in a fetus with normally functioning autonomic nervous system.

Accelerations of 15 beats per minute above baseline and for 15 seconds from

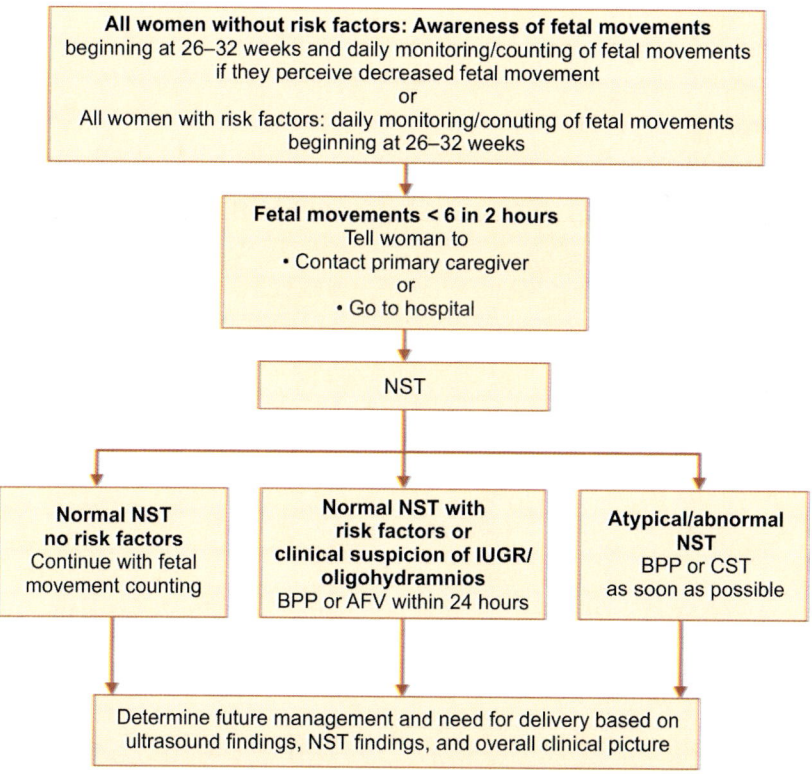

Fig. 7.1: Algorithm depicting use of fetal movement count.[2]

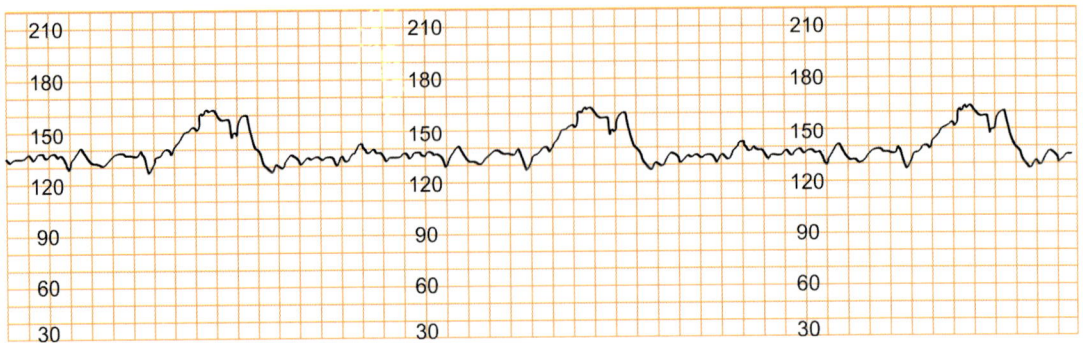

Fig. 7.2: Reactive NST

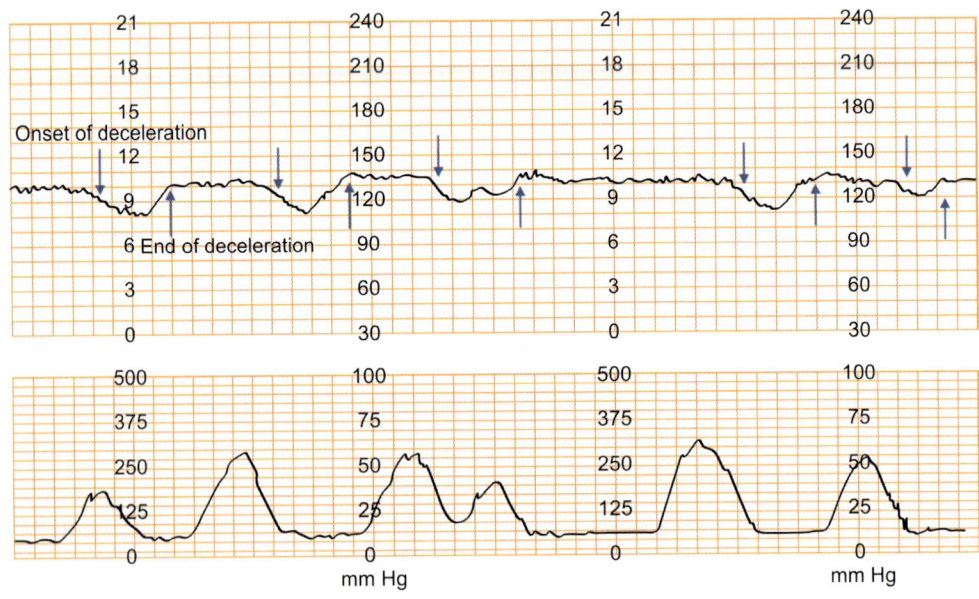

Fig. 7.3: Late Decelerations

the baseline in a 20 to 40 minute period are considered reactive as depicted in Fig. 7.2 and are a measure that has stood the test of time as a predictor of fetal well-being at that point in time. However, fetal decelerations as depicted in Fig. 7.3 as associated with fetal distress.

Despite widespread use, there is little evidence that antenatal non-stress test can reduce perinatal morbidity or mortality in low-risk women. It has definitely found to be beneficial in women with risk factors for adverse perinatal outcome.

In most cases a normal NST is predictive of good perinatal outcome for one week (providing the maternal-fetal condition remains stable), except in women with insulin dependent diabetes or with a postdates pregnancy, in which case NSTs are recommended at least twice weekly. The various categories of NST are classified in Table 7.2.

The negative predictive value of NST alone for predicting stillbirth within 1 week of a normal test is 99.8%; for BPP, modified BPP, and CST, it is greater than 99.9%.[10]

Table 7.2: Classification of NST

Parameter	Normal NST (Previously reactive)	Atypical NST (Previously non-reactive)	Abnormal NST (Previously non-reactive)
Baseline	110–160 bpm	• 100–110 bpm • > 160 bpm < 30 minutes • Rising baseline	• Bradycardia <100 bpm • Tachycardia > 160 for > 30 minutes • Erratic baseline
Variability	• 6–25 bpm (moderate) • ≤ 5 (absent or minimal) for < 40 minutes	≤ 5 (absent or minimal) for 40–80 minutes	• ≤ 5 for ≥ 80 minutes • ≥ 25 bpm >10 minutes • Sinusoidal
Decelerations	None or occasional variable < 30 seconds	Variable decelerations 30–60 seconds duration	• Variable decelerations > 60 seconds duration • Late deceleration(s)
Accelerations term fetus	≥ 2 accelerations with acme of ≥ 15 bpm, lasting 15 seconds < 40 minutes of testing	≤ 2 accelerations with acme of ≥ 15 bpm, lasting 15 seconds in 40–80 minutes of testing	≤ 2 accelerations with acme of ≥ 15 bpm, lasting 15 seconds in > 80 minutes
Preterm fetus (<32 weeks)	≥ 2 accelerations with acme of ≥ 10 bpm, lasting 10 seconds < 40 minutes of testing	≤ 2 accelerations of ≥ 10 bpm, lasting 10 secconds in 40–80 minutes	≤ 2 accelerations of ≥ 10 bpm, lasting 10 seconds in > 80 minutes
Action	**Further assessment optional** based on total clinical picture	**Further assessment required**	**Urgent action required** An overall assessment of the situation and further investigation with U/S or BPP is required. Some situations will require delivery

Recommendations for NST[2]

- Antepartum non-stress testing may be considered when risk factors for adverse perinatal outcome are present. (III-B)
- In the presence of a normal non-stress test, usual fetal movement patterns, and absence of suspected oligohydramnios, it is not necessary to conduct a biophysical profile or contraction stress test. (III-B)
- A normal non-stress test should be classified and documented by an appropriately trained and designated individual as soon as possible (ideally within 24 hours).

2. Contraction Stress Test (CST)

The Contraction Stress Test (CST), or Oxytocin Challenge test, is a test of fetal well-being first described by Ray et al. in 1972. It evaluates the response of the FHR to induced contractions and was designed to unmask poor placental function.

The CST may be performed using maternal nipple stimulation or an oxytocin infusion.

A CST is considered *positive* if late decelerations occur with more than 50% of the induced contractions (even if the goal of three contractions in 10 minutes has not yet been reached). A *negative* CST has a normal baseline FHR tracing without late decelerations. An equivocal test is defined as repetitive decelerations, not late in timing or pattern. A CST is deemed unsatisfactory if the desired number and length of contractions is not achieved or if the quality of the cardiotocography tracing is poor.

In recent times when uteroplacental function is often evaluated by biophysical variables (e.g. biophysical profile) or vascular flow measurements (e.g. Doppler interrogation of uterine or fetal vessels), the contraction stress test is now being performed much less often.

The CST has a high negative predictive value (99.8%).[11] Its positive predictive value for perinatal morbidity however is poor (8.7–14.9%). It should never be used alone to guide clinical action. The corrected perinatal mortality rate within one week of a negative contraction stress test is 1.2/1000 births.[2]

Recommendations on contraction stress test:[2]

1. The contraction stress test should be considered in the presence of an atypical non-stress test as a proxy for the adequacy of intrapartum uteroplacental function and, together with the clinical circumstances, will aid in decision-making about timing and mode of delivery (III-B).
2. The contraction stress test should *not* be performed when vaginal delivery is contraindicated (III-B).
3. The contraction stress test should be performed in a setting where emergency cesarean section is available (III-B).

3. Biophysical Profile (BPP)

The BPP is an evaluation of current fetal well-being. It is performed over 30 minutes and assesses fetal behaviour by observing fetal breathing movement, body movement, tone, and amniotic fluid volume. In the presence of intact membranes, functioning fetal kidneys, and unobstructed urinary tract, decreased amniotic fluid reflects decreased renal filtration due to redistribution of cardiac output away from the fetal kidneys in response to chronic hypoxia. The components of the fetal biophysical profile that are usually evaluated as enlisted in Table 7.3.

Each of these individual ultrasound assessed variables is scored 0 (if absent) or 2 (if present) and summed for a maximum score of 8. The inclusion of the NST brings the maximum possible score to 10 when the NST is normal.

Table 7.3: Components of fetal biophysical profile[12]	
Component	**Definition**
Fetal movements	3 body or limb movements
Fetal tone	One episode of active extension and flexion of the limbs; opening and closing of hand
Fetal breathing movement	episode of ≥ 30 seconds in 30 minutes Hiccups are considered breathing activity.
Amniotic fluid volume	Single 2 × 2 cm pocket is considered adequate.
Non-stress test	2 accelerations >15 beat per minute of at least 15 seconds duration.

The original BPP included all five components in every pregnancy assessment. A more recent approach is to carry out the ultrasound components, reserving the NST for pregnancies in which one of the ultrasound components is absent.

A score of 10 or 8 (including 2 for fluid present) is considered normal, 6 is considered equivocal, and 4 or less is abnormal. Reassessment of a patient with an equivocal result of 6 of 10 with normal fluid will be reassuring in 75% of cases.

Some centres carry-out a "modified" BPP as the primary screening test of antenatal surveillance. The modified BPP consists of a non-stress test and an AFI (>5 cm is considered adequate). If either assessment measure is of concern, then the complete BPP is performed. The perinatal mortality rates correlating with the BPP scores are listed in Table 7.4.

Table 7.4: Perinatal mortality and the biophysical profile score	
Score	**Perinatal mortality/1000**
8–10	1.86
6	9.76
4	26.3
2	94.0
0	285.7

Recommendation: Biophysical Profile[2]

1. In pregnancies at increased risk for adverse perinatal outcome and where facilities and expertise exist, biophysical profile is recommended for evaluation of fetal well-being (I-A).
2. When an abnormal biophysical profile is obtained, the responsible physician should be informed immediately. Further management will be then determined by the overall clinical situation (III-B).

4. Uterine artery Doppler

Doppler ultrasound of the uterine arteries is a noninvasive method of assessing the resistance of uterine vessels supplying the placenta. In normal pregnancies, there is an increase in blood flow velocity and a decrease in resistance to flow, reflecting the transformation of the spiral arteries. In pregnancies complicated by hypertensive disorders, Doppler ultrasound of the uterine artery shows increased resistance to flow, early diastolic notching, and decreased diastolic flow. A positive uterine artery Doppler screen consists of mean resistance index of >0.57, pulsatility index >95th centile, and/or the presence of uterine artery notching.

Indications for Uterine Artery Doppler at 17 to 22 Weeks[2]

Previous obstetrical history

- Previous early onset gestational hypertension
- Placental abruption
- Intrauterine growth restriction
- Stillbirth.

Risk factors in current pregnancy

- Pre-existing hypertension
- Gestational hypertension
- Pre-existing renal disease
- Longstanding Type I diabetes with vascular complications, nephropathy, retinopathy
- Abnormal maternal serum screening
- Low PAPP-A (consult provincial lab for norms).

Approximately 1% of at-risk pregnancies have abnormal uterine artery Doppler resistance and/or notching after 26 weeks of gestation. The likelihood of development of gestational hypertension and/or growth restriction in these pregnancies is increased fourfold to eightfold.

Conversely, normal uterine artery pulsatility index or resistance index significantly reduces the likelihood of these pregnancy complications (negative predictive value varying between 80 and 99%). Data on the use of uterine artery Doppler screening in populations without risk factors for adverse outcome is less well-substantiated.

Figures 7.4 and 7.5 highlight normal and abnormal uterine artery Doppler waveforms respectively.

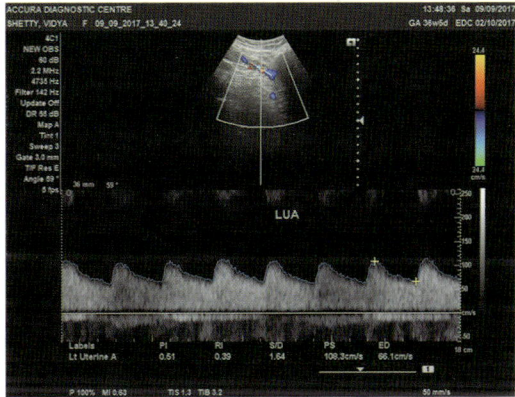

Fig. 7.4: Normal uterine artery Doppler waveform.

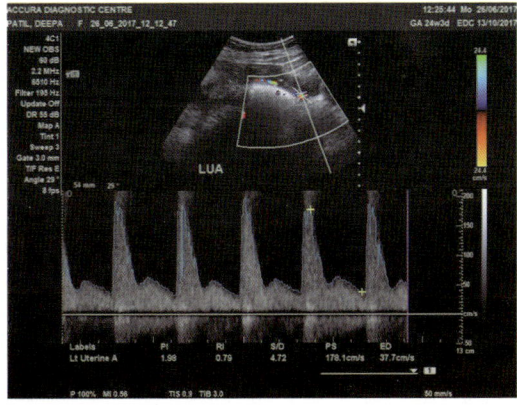

Fig. 7.5: Abnormal uterine artery Doppler waveforms displaying notching.

Recommendations: Uterine Artery Doppler[2]

1. Where facilities and expertise exist, uterine artery Doppler may be performed at the time of the 17 to 22 weeks' gestation and detailed anatomical ultrasound scan in women with the presence of factors for adverse perinatal outcome (II-A).

2. Women with a positive uterine artery Doppler screen should have the following:
   - A double marker screen (for alpha feto-protein and beta hCG) if at or before 18 weeks' gestation (III-C).
   - A second uterine artery Doppler at 24 to 26 weeks. If the uterine artery Doppler is positive at the second scan, the woman should be referred to a maternal-fetal medicine specialist for management (III-C).

5. Umbilical Artery Doppler

Increased resistance to forward flow in the umbilical circulation is characterized by abnormal systolic to diastolic ratio, Pulsatility Index (PI) or Resistance Index (RI) greater than the 95th centile and implies decreased functioning vascular units within the placenta.

The Cochrane meta-analysis[13] of randomized trials on the use of umbilical artery Doppler in pregnancies with risk factors for adverse perinatal outcome demonstrates a clear reduction in perinatal mortality in normally formed fetuses. This form of fetal surveillance has been shown to improve perinatal mortality in randomized controlled trials.

Figures 7.6 and 7.7 highlight normal and abnormal umbilical artery Doppler waveforms respectively.

Recommendations: Umbilical Artery Doppler[2]

1. Umbilical artery Doppler should not be used as a screening tool in healthy pregnancies, as it has not been shown to be of value in this group (I-A).

2. Umbilical artery Doppler should be available for assessment of the fetal placental

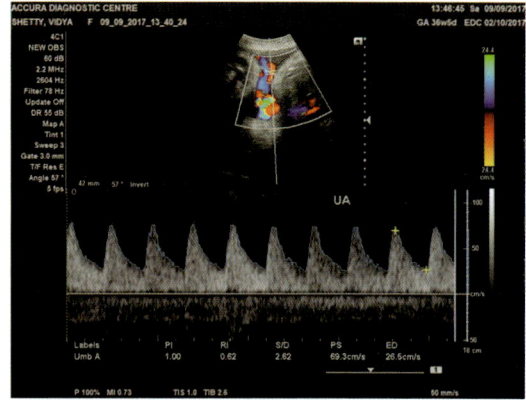

Fig. 7.6: Normal umbilical artery Doppler waveforms showing good end diastolic flow.

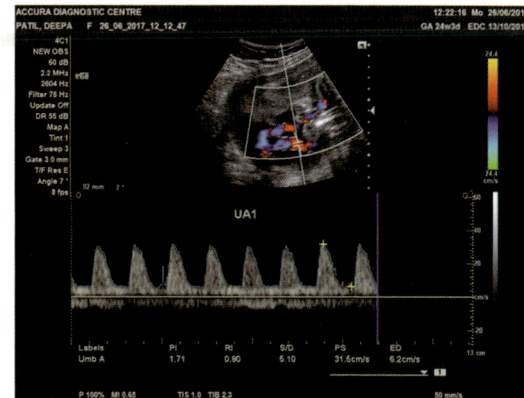

Fig. 7.7: Abnormal umbilical artery Doppler waveforms showing reduced end diastolic flow.

circulation in pregnant women with suspected placental insufficiency (I-A).

Fetal umbilical artery Doppler assessment should be considered at time of referral for suspected growth restriction, or during follow-up for suspected placental pathology.

3. Depending on other clinical factors, reduced/absent/reversed umbilical artery end-diastolic flow is an indication for enhanced fetal surveillance or delivery. If delivery is delayed to improve fetal lung maturity with maternal administration of glucocorticoids, intensive fetal surveillance until delivery is suggested for those fetuses with reversed end-diastolic flow (II-1B).

6. Other Fetal Artery Doppler

Other fetal artery Doppler parameters when Doppler expertise is available progression of cardiovascular compromise in the fetus with intrauterine growth restriction.

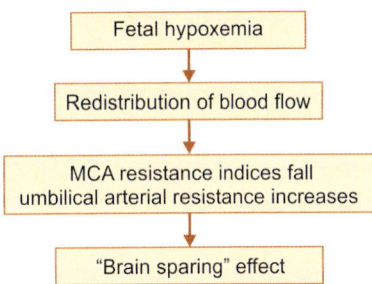

Decreased cerebral impedance, like descending aorta impedance also leads to reversal of blood flow in the aortic isthmus. Changes in the cerebral flow parameters, however, do not correlate well with the final stages of asphyxia or compromise and therefore are not helpful in choosing timing for delivery. Increased resistance in the umbilical arteries and descending aorta does lead, however, in an increase in right ventricular end-diastolic pressure (after load), leading to decreased right ventricular compliance and increased venous pressure in the right atrium and systemic veins. This can be detected using transtricuspid E/A (early and late diastolic filling) ratios, which increase with decreased ventricular compliance.

Further deterioration of right ventricular contractility will lead to right ventricular dilatation and tricuspid regurgitation (insufficiency), further exacerbating right atrial filling pressure and resistance to venous filling. Resistance to venous filling is reflected best by increased pulsatility in the ductus venosus during atrial contraction, a finding highly correlated with impending asphyxia and acidosis. Further increases in systemic venous pressures lead to maximum dilatation of the ductus venosus and direct transmission of cardiac impulses to the umbilical vein, causing umbilical venous pulsations. This finding is shown to be highly correlated with severe acidosis and impending fetal demise.

Middle Cerebral Artery Peak Systolic Velocity as a Predictor of Fetal Anemia

- MCA PSV is highly correlated with severe fetal anemia (sensitivity as high as 100%)[2] in fetuses with non-immune hydrops or when prospectively following a fetus at risk of parvovirus B19-induced fetal anemia,
- MCA PSV serves as a useful measure of fetal anemia severe enough to require IUT.

Figure 7.8 highlights abnormal middle cerebral artery Doppler waveform.

Figure 7.9 summarises the use of Doppler in antenatal monitoring.

When to begin antenatal testing and how frequent?

Prognosis for neonatal survival is the most important consideration followed by maternal high-risk factors. Most authorities recommend to start testing by 32 weeks. Pregnancies with severe complications might require testing as early as 26 to 28 weeks. The frequency of testing is arbitrarily set at 7 days.

ACOG Guidelines at a Glance: Antepartum Fetal Surveillance[10]

- A warning sign that a fetus may be at risk of compromise is maternal perception of

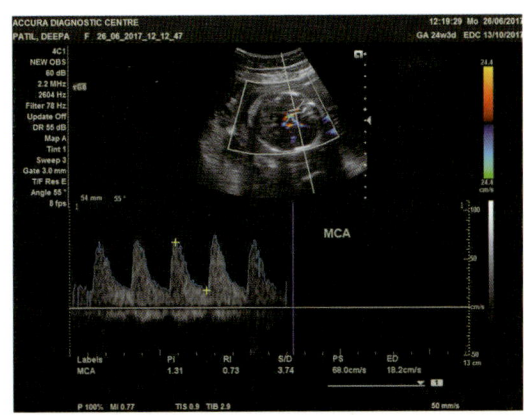

Fig. 7.8: Abnormal middle cerebral artery Doppler waveform (MCA/UmbA S/D ratio <1).

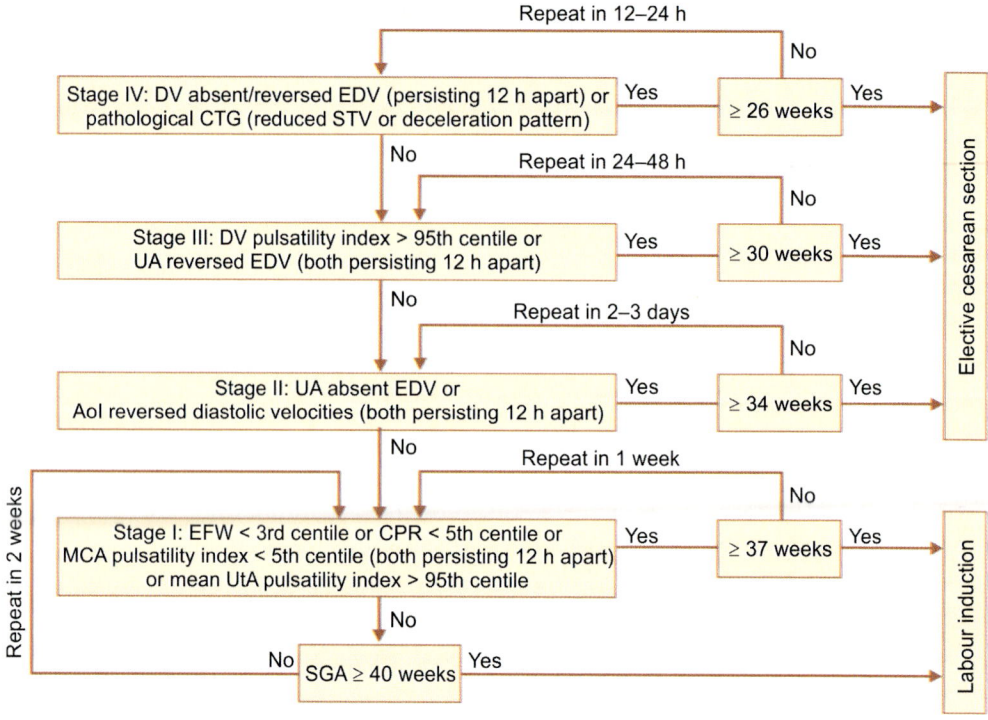

Fig. 7.9: Antenatal monitoring with Doppler studies.[2]

decrease in fetal movement. A nonreassuring count should prompt notification for further fetal assessment.

- The NST and the ultrasound biophysical profile (BPP) are the primary antenatal fetal surveillance methods. A normal BPP score along with a reactive NST is an indication of fetal well-being.

- With the wider usage of umbilical artery Doppler velocimetry, particularly in the surveillance of Fetal Growth Restriction (FGR), the Contraction Stress Test (CST) is now rarely used to assess for fetal compromise or potential hypoxemia. There is no evidence that inclusion of umbilical artery Doppler velocimetry in antenatal surveillance provides additional benefit in the assessment of a normally growing fetus.

- The negative predictive value of NST alone for predicting stillbirth within 1 week of a normal test is 99.8%; for BPP, modified BPP, and CST, it is greater than 99.9%.

- Antenatal fetal surveillance has stood the test of time with regard to the goal of preventing stillbirth in the fetus at risk based on indications for testing. The additional cost for testing in the appropriate setting appears to have benefit. However, clinicians should be reminded that the least costly antenatal surveillance modality is maternal fetal movement assessment as a test for well-being in low- and high-risk women, even if its effectiveness in preventing stillbirth is uncertain.

REFERENCES

1. American Academy of Paediatrics and American College of Obstetricians and Gynecologists: Guidelines for prenatal care, 7th ed. Washington, 2012.

2. Robert Liston, DianeSawchuck, David Young. SOGC Clinical Practice Guideline, Journal of Obstetrics and Gynaecology Canada, 2007; 29(9):s9

3. Robert PJ, HoJJ, Valliapan j, Sivasangari S. Measuring the height of the uterus from the symphysis pubis (SFH) in pregnancy for detecting problems with fetalgrowth. **Cochrane Summaries.** 2012. Avail at http://summaries. cochrane.org/CD008136/measuring-the-height-of-the-uterus-from-the-symphysis-pubis-sfhin-pregnancy-for-detecting-problems-with-fetal-growth.

4. American College of Obstetricians and Gyne-cologists: Fetal growth restriction. Practice Bulletin No. 134, May 2013b.

5. Froen JF. A kick from within—fetal movement counting and the cancelled progress in antenatal care. J Perinat Med 2004;32(1):13–24.

6. Grant A, Elbourne D, Valentin L, Alexander S. Routine formal fetal movement counting and risk of antepartum late death in normally formed single tons. Lancet 1989;2(8659):345–9.

7. Pearson JF, Weaver JB. Fetal activity and fetal wellbeing: an evaluation. BrMed J 1976; 1(6021): 1305–7.

8. Moore TR, Piacquadio K. A prospective evalua-tion of fetal movement screening to reduce the incidence of antepartum fetal death. Am J Obstet Gynecol 1989;160(5 Pt 1):1075–80.

9. Sadovsky E, Weinstein D, Even Y. Antepartum fetal evaluation by assessment of fetal heart rate and fetal movements. Int J Gynaecol Obstet 1981; 19(1):21–6.

10. American College of Obstetricians and Gynae-cologists Practice Bulletin number 145: Antenatal Fetal Surveillance. Obstet Gynecol 2014;124: 182–92.

11. Freeman RK, Anderson G, Dorchester W. A pros-pective multi-institutional study of antepartum fetal heart rate monitoring. II. Contraction stress test versus nonstress test for primary surveillance. Am J Obstet Gynecol 1982;143(7):778–81.

12. Manning FA. Dynamic ultrasound-based fetal assessment: the fetal biophysical profile score. Clin Obstet Gynecol 1995;38(1):26–44.

13. Neilson JP, Alfirevic Z. Doppler ultrasound for fetal assessment in high risk pregnancies [Cochrane review]. In: Cochrane Database of Systematic Reviews 1996 Issue 4. Chichester (UK): John Wiley & Sons, Ltd; 1996. DOI: 10.1002/14651858.CD000073.

Intranatal Monitoring and Use of Fetal Scalp Monitor

Reva Tripathi, Meenoo S

Perinatal Mortality (PNM) is a major problem in developing countries and is an important contributor to infant mortality rate of about 41 per 1000 live births. One of the chief causes of PNM is intrapartum stillbirth. This has declined worldwide by 14.5%, from 22.1 in 1995 to 18.9 stillbirths per 1000 births in 2009 though, South Asia and sub-Saharan Africa contributed majorly to about 76.2% of stillbirths.[1] Perinatal morbidity includes seizures and encephalopathy occurring in the neonatal period which may be due to a variety of reasons but is commonly attributed to intrapartum problems. In order to decrease intrapartum insults, some type of intrapartum fetal monitoring is essential as uterine contractions cause decrease in uteroplacental blood flow which results in reduced oxygen delivery to the fetus. Most normal fetuses can withstand this decrease in blood flow however, high-risk fetuses may not tolerate this reduction and go into a state of fetal hypoxia, followed by asphyxia (hypoxic acidemia), hypercapnia, and metabolic acidosis. Intrapartum monitoring is aimed at minimising this pathology in an attempt to decrease problems of long term neuro developmental delay.

Methods of Fetal Monitoring
Intermittent ausculation   Electronic fetal monitoring
External   Internal

INTERMITTENT AUSCULTATION (IA)

Intermittent auscultation of Fetal Heart Rate (FHR) is an accepted method of intrapartum fetal monitoring in women with spontaneous labour in the absence of risk factors. Intermittent auscultation (IA) can be performed using Pinard's or Laennac stethoscope or hand held Doppler. It should be performed for 60 seconds, which should be started just before contraction, heard throughout the duration and continued after completion of the contraction to ensure that FHR is within normal range. The frequency of IA varies according to the stage of labour (Table 8.1). One must be cautious to differentiate between maternal pulse and FHR to avoid unnecessary interventions. If the fetal heart is technically inaudible so that the FHR cannot be established, then electronic fetal monitoring should be commenced. IA method of monitoring has a lower intervention rate without compromising neonatal outcome.[2]

Advantages
- Less costly.
- Can be done in any hospital.
- Patients can be ambulatory and can take posture of their choice.

Disadvantages
- Ideally requires one to one monitoring, i.e. one doctor/midwife for one patient which is not likely to be possible in many centres.

Table 8.1: Recommended frequency of auscultation		
Recommending agency	Frequency in active phase of first stage	Frequency in second stage
AWHONN (2000)	15–30 minutes	5–15 minutes
RCOG (2001)	15 minutes	5 minutes
ACOG (2005)	15 minutes	5 minutes
SOGC (2007)	15–30 minutes	5 minutes

- Minor or subtle changes cannot be reliably assessed by IA.
- Bradycardia may not be detected.
- No records are available for subsequent analysis.
- Subjectivity may be high.

ELECTRONIC FETAL MONITORING (EFM)

EFM may be performed with an external or internal monitor. However, conventionally it is done by external method.

External Monitoring

External monitoring has two components, i.e. FHR monitoring and recording of uterine contractions. It is the most commonly used method of fetal monitoring after intermittent auscultation. This is a form of electronic fetal surveillance that has an ultrasound transducer which detects Doppler shift and interprets the FHR using an inbuilt computer software.

Indications: It isused mostly in high-risk conditions like fetal growth restriction, abruption, oligohydramnios, etc. The indications are discussed in detail elsewhere in this book.

When to start EFM?

Each patient must be individualised and decision to start the monitoring must be taken by the treating obstetrician. It is commonly done in all high-risk patients after the patient enters active phase of labour. However in places where provisions are not sufficient, electronic fetal monitoring may be done for short intervals every few hours and the remaining time monitoring can be done by IA unless otherwise indicated.

Admission CTG: Admission CTG was first described by Ingemarrson and Arulkumaran in 1986 as a 20 minute CTG performed on admission for delivery. It was used as a screening test to identify the fetus at risk of intrapartum hypoxia and acidosis and also to triage the women who are in need of continuous fetal electronic monitoring. Most patients undergoing this test generally would be in latent phase or in early part of active phase.

Meta-analyses of controlled trials found that women randomized to the labour admission test were more likely to have minor obstetric interventions like epidural analgesia [relative risk (RR 1.2), continuous electronic fetal monitoring (RR 1.3)] and fetal blood sampling (RR 1.3) compared with women randomized to auscultation on admission. There were no significant differences in any of the other outcomes.[3] From observational studies, prognostic value for various outcomes was found to be generally poor. There is no evidence supporting labour admission test being beneficial in low-risk women without risk factors for adverse perinatal outcome.[5] However, fetal heart tracings on admission are recommended for women with risk factors for adverse perinatal outcome[4] (III-B).

Evidence suggests that compared with IA, continuous electronic fetal monitoring in labour is associated with an increase in the rates of cesarean sections and instrumental vaginal births. Continuous electronic fetal monitoring during labour is associated with a reduction in neonatal seizures but with no significant differences in cerebral palsy, infant mortality, or other standard measures of neonatal well-being like Apgar scores and cord-blood gases.[5]

Advantages of external monitoring

- It is noninvasive
- Cervical dilatation or rupture of membranes is not necessary.

Disadvantages of external monitoring

- Likely to require readjustment with maternal or fetal movements.
- Difficult to obtain a clear tracing in obese women or those with polyhydramnios.
- More prone to signal loss and artefacts.

Internal Monitoring

It is rarely used in clinical practice because of its invasive nature and minimal advantages over external monitoring.

There are two types of internal monitoring for the fetus, one is to record FHR and the other is to record fetal ECG. The FHR analysis is similar to the external one whereas the fetal ECG analyses the time intervals between successive cardiac cycles and therefore measures ventricular depolarization cycles. The difference in analysis depends on the software in the machine.

Internal monitoring is done using a fetal scalp electrode (Fig. 8.1). The fetal scalp electrode consists of a single ECG electrode in the form of a spiral needle which is attached to paired wires that extend out of the mother's vagina (Fig. 8.2). These wires are inserted into a ground plate that is placed on the maternal thigh. The ground plate transmits the signal to the monitor through a cable. The electrode is placed within an introducer and inserted so that the mother's vagina is protected from the needle. Internal monitoring may also be done for evaluation of uterine contractions.

Procedure

Initially, patient must be explained about her status and the need for internal monitoring and consent must be taken. The leg plate is attached to the maternal thigh. Per vaginal examination under aseptic precautions is performed and membranes are ruptured, if not already done. Fetal position is assessed and a location over parietal or occipital bone, away from sutures, is chosen for application of electrode. In cases of trial breech delivery, application to the fetal buttocks may also be considered.

The electrode along with the introducer is inserted into the vagina up to the fetal scalp.

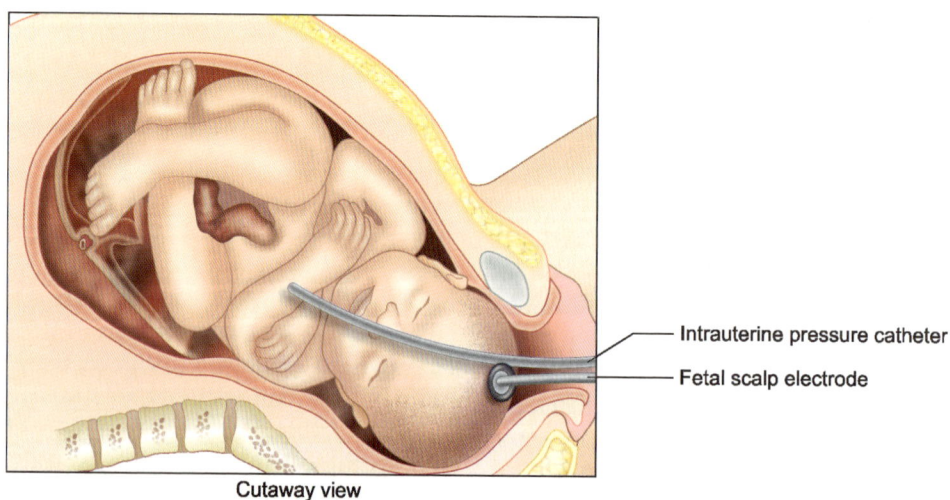

Intrauterine pressure catheter

Fetal scalp electrode

Cutaway view

Fig. 8.1: Application of fetal scalp electrode.

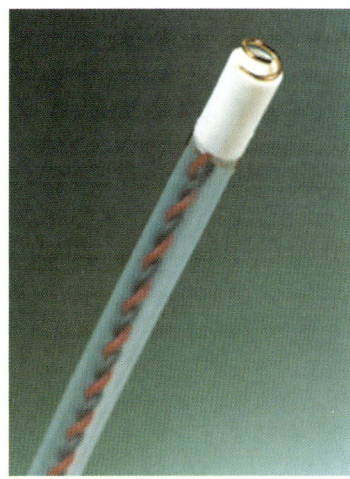

Fig. 8.2: Fetal scalp electrode.

This is done such that the needle is not exposed. Then the electrode is cautiously turned clockwise so that head is not disengaged which may lead to cord prolapse. Then the wires and monitor are connected and tracings are evaluated. For removal before cesarean or delivery, the electrode must be rotated counter clockwise so as to safely detach the needle from the fetal scalp.

Advantages
- More accurate than external monitoring.

Disadvantages
- One should be cautious about the presenting part of the fetus so that electrode is not placed on delicate structures such as sutures and fontanels.
- It requires rupture of membranes.
- There is increased risk of vertical transmission of infections and hence is contraindicated in patients with active genital herpes infection, Hepatitis B, C, D and E and HIV.
- It should be avoided in preterm fetuses because of risks of intracranial hemorrhage.

Internal monitoring has theoretical advantage of accuracy over external monitoring but due to its invasive nature, external fetal monitoring is commonly used.

Uterine contractions: These are bell-shaped gradual increases in the uterine activity signal followed by roughly symmetric decreases, with 45–120 seconds in total duration. Contractions can be evaluated either by tocodynamometer externally or intrauterine pressure catheter. Uterine contractions have also been recorded by electromyography. Hadar et al. in 2014 showed that electromyography was not inferior to tocodynamometer in interpretability of tracings, frequency of contractions and timing of contraction.[6]

Recording of contractions is important because FHR changes have to be correlated with uterine contractions. Decelerations are benign if occur during contractions whereas may need intervention in case they occur after contraction has ceased.

SPECIAL SITUATIONS

Monitoring of Twins

Continuous external fetal monitoring of twin fetuses should be preferably done with dual channel monitors so that simultaneous monitoring of both FHR is possible. Sometimes duplicate monitoring of the same twin might occur and this can be identified by identical tracings. Some monitors are provisioned with an alarm when this situation is suspected. During second stage of labor, it is preferable to monitor the first twin by internal monitoring and the second twin by external monitoring.

Monitoring of Trial of Labour After Cesarean Section (TOLAC)

Continuous fetal monitoring is recommended in patients undergoing trial of labour after cesarean section.[7] The main complication that can occur in a patient undergoing TOLAC is uterine rupture though it is less than 1%. Electronic monitoring is generally preferred in these patients as loss of variability or subtle fetal heart changes may be the first sign of imminent scar dehiscence followed by the appearance of decelerations. A good CTG trace is a valuable contribution to ensure VBAC.

Monitoring in Water Birth

In water birth, FHR is auscultated intermittently with stethoscope or waterproof Doppler. During this auscultation, patient is asked to raise the buttocks out of water and if patient is not able to raise the buttocks, then the midwife or doctor can wear long gloves and use the Doppler probe underwater. If any abnormal FHR is found, then patient is brought out of the water tub and monitored with an electronic fetal monitor. If the CTG tracing is normal, then patient can re-enter the tub. In case of abnormal tracings, patient is to be managed as a normal routine labouring patient. Electronic monitoring is avoided in submerged patients because of the potential risk of electric shock despite use of precautions.

MANAGEMENT AFTER DETECTION OF INTRA-PARTUM FETAL HEART RATE ABNORMALITIES

Initial uterine resuscitation is done which is described elsewhere in this book. After these measures, other techniques that can be considered are digital fetal scalp stimulation, fetal blood sampling and fetal ECG.

Digital Fetal Scalp Stimulation

Digital fetal scalp stimulation is an indirect assessment of acid–base status. It is done by stroking the scalp during a per vaginal examination, where consequent to the stimulation an acceleratory response is induced due to sympathetic activation which indicates a normal fetus. It must be realized that although an acceleratory response is consistent with a reasonable likelihood of fetal well-being, the absence of this response does not necessarily predict fetal compromise. However, when there is a lack of acceleration, further assessment is necessary, preferably by fetal scalp blood sampling. The technique is important,

because using excessive pressure may produce vagal bradycardia and should therefore be avoided. It is best to avoid digital scalp stimulation during a deceleration, as the deceleration reflects a vagal response and that would prevent any sympathetic nerve response during scalp stimulation. This should be done in between contractions and when FHR is at a stable condition.

Fetal Blood Sampling (FBS)

It is indicated in patients with abnormal intrapartum CTG tracings after 34 weeks period of gestation and to decide if delivery is required whether by cesarean section or operative vaginal delivery.

Procedure

The patient's condition is explained and consent for the procedure must be taken. It is preferable to do it in left lateral position to avoid aortocaval compression. Sterile vaginal examination is done using anamnioscope which is passed into the vagina and positioned against the fetal head. The amnioscope should be placed away from caput or the fontanels. The fetal scalp is cleaned with swab sticks or dry cotton wool using sponge holding forceps. An assistant sprays skin coolant down the amnioscope to the area where the blood sample is to be obtained for 3 seconds and then after 30 seconds, hyperemia occurs. The fetal scalp blade is held firmly between the fingers and thumb and firm pressure is applied to the fetal scalp to make a tiny incision with the blade. If no blood is withdrawn, then caput formation must be excluded, and ensured that pressure applied is constant. A droplet of blood is allowed to form on the scalp and a heparinised capillary tube is used to collect samples. The capillary tube is shaken gently and finally; column collected in the tube should be 20–25 mm without any bubbles. Estimation of lactate requires only 5 µL whereas, pH analysis requires 30 to 50 µL. If sufficient sample is available for analysis, then lactate and pH levels both may be done.

Table 8.2: Classification of fetal scalp blood sampling results		
pH	**Lactate**	**Intervention**
> 7.25	< 4.2 mmol/L	FBS should be repeated if the FHR abnormality persists.
7.21–7.24	1.2–4.8 mmol/L	Repeat FBS within 30 minutes or consider delivery if rapid fall since last sample.
< 7.20	> 4.8 mmol/L	Immediate delivery indicated.

However, technology for analysis of pH is more commonly available and local situation will decide which analysis is to be done (Table 8.2).

Contraindications

- Family history of hemophilia or suspected fetal bleeding disorder (eg. fetal thrombocytopenia)
- Face presentation
- Maternal infection (HIV, viral hepatitis, herpes simplex, suspected intrauterine sepsis).

Limitations

- Skill and experience of the operator.
- Need for cervical dilatation.
- Maternal discomfort.
- May need to perform multiple FBS.

Complications

- Inadvertent Laceration of maternal cervix or vagina.
- Increased chances of endometritis.
- Fetal scalp injury which some times may lead to CSF leak if fontanels are involved.
- Fetal ocular penetrating injury.
- Increased chance of neonatal scalp abscesses due to *E. coli, gonococcus* or Group A *Streptococci.*

Cochrane review 2015 showed that Fetal Blood Sampling for Lactate testing was more successful than pH testing as it requires less blood. However, there was no difference in newborn outcomes, including the number of babies with low Apgar scores, low pH in their cord blood or admissions to the neonatal intensive care nursery. There were also no differences in the cesarean sections rate, forceps or vacuum deliveries between the two groups.[8]

NEWER TECHNIQUES FOR INTRAPARTUM DIAGNOSIS OF FETAL RISK/JEOPARDY

Fetal Pulse Oximetry (FPO): The development of reflectance pulse oximetry has made it possible to measure fetal oxygen saturation during labour. Intrapartum reflectance pulse oximetry measures oxygen saturation in capillary area by detecting pulsatile signals of the fetal vessels.

A sensor is placed transvaginally through the cervix over fetal cheek or temple. This can be done only when there is at least 2 cm dilatation with ruptured membranes and fetus in vertex positon.

Cochrane review 2014 showed fetal pulse oximetry together with CTG showed no difference in cesarean section rates overall, or any difference in the maternal and neonatal outcomes, compared with CTG alone. However, its use reduced cesarean sections done for fetal distress.[9]

Fetal pulse oximetry may help to decrease the number of fetal blood sampling and the incidence of operative interventions performed.[4]

There is no substantial evidence currently to recommend the use of FPO as an independent method of electronic fetal surveillance as it is used mainly in research settings though it is less invasive and has the advantage of avoiding sampling of fetal blood.

FETAL ECG ANALYSIS

Fetal ECG analysis is a type of internal monitoring though not commonly used in clinical practice but mostly used in research settings. The use of this method in fetal surveillance is based on the pathophysiology of myocardial response to hypoxia which is an extrapolation of changes that take place in adult myocardium when ischemia or infarction occurs.

ST analysis (STAN) is a system for fetal surveillance that displays the FHR and information resulting from the computerised analysis of ST interval of the fetal ECG (Figs 8.3A and B). This is based on ECG changes determined by myocardial adaptation to oxygen deficiency. The fetal heart ability to pump blood is dependent on a balance between myocardial workload and metabolic reactions. To sustain the workload, energy is produced by these cells using aerobic metabolism in a normal healthy fetus where the ECG shows a normal ST-waveform.

In case of oxygen insufficiency, a fetus with intact defence mechanisms reacts by releasing stress hormones and switches from aerobic to anaerobic metabolism to sustain the workload. This anaerobic metabolism in the myocardial cells causes changes in the ST interval of the fetal ECG. The ST interval consists of the ST segment and the T-wave. A normal ST interval in fetal ECG is characterized by a horizontal or upward sloping ST segment and a T-wave with constant and stable amplitude (Fig. 8.4).

In anaerobic metabolism, the breakdown of glycogen initiates the release of potassium ions, which causes an increase in the height of T-wave. The T-wave rise is an indirect measurement of utilisation of glycogen in the heart muscle cells. ST depression and negative T-waves are other significant ST changes that indicate a compromised myocardial performance.

The STAN machine has a monitor showing both the fetal CTG and analysis of fetal ECG continuously. It is done using fetal scalp electrode similar to that used during internal CTG monitoring. A single spiral fetal scalp electrode records the signal necessary for ST analysis and CTG simultaneously.

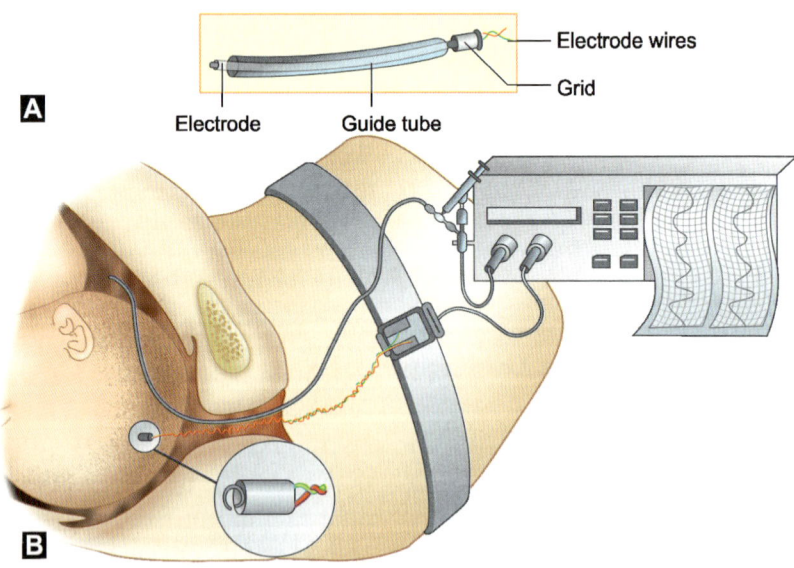

Figs 8.3A and B: Fetal ECG monitor.

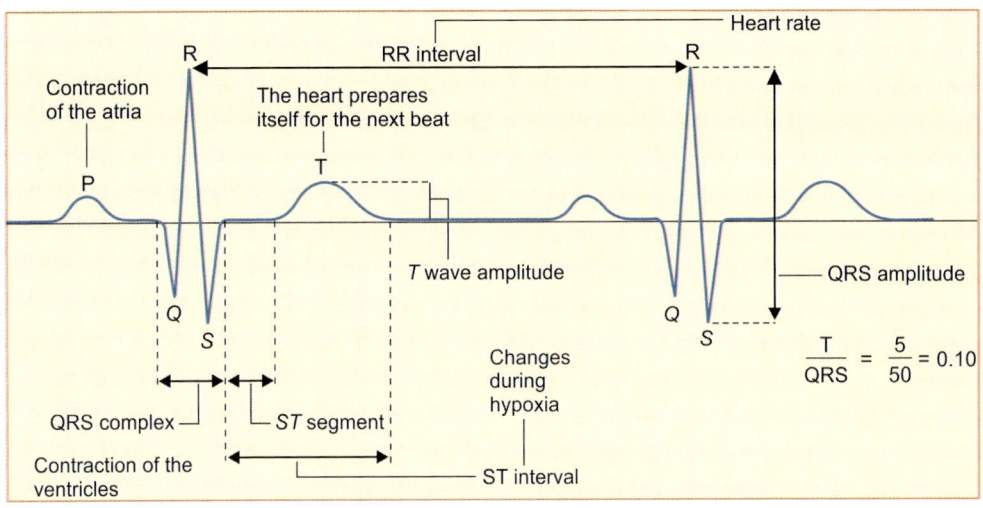

Fig. 8.4: Fetal ECG complex.

The fetal ECG signal is automatically recorded, processed and analysed by the STAN system computerized software and results are displayed on the screen. When the monitor has detected any significant change in the ST interval, these are displayed as "ST events" on the main screen. These ST events are considered to be pathological but their analysis is used as an adjunct to CTG monitoring to determine whether obstetrical intervention is needed or not.

STAN guidelines recommend that clinical action must be done based on the combined assessment of the CTG classification and ST events as they mainly represent the severity of hypoxia when combined with CTG.[10]

When the fetus is connected to the STAN Machine it calculates the normal baseline T/QRS ratio for the individual fetus. This is done by analysing the fetal ECG complexes that are received through the fetal scalp electrode. If there is good signal quality, this takes approximately four to five minutes. Once this is over, the computer records this value as the 'normal baseline' for the individual fetus in question. Subsequently, the computer analyses every 30 fetal ECG complexes and compares with the original 'baseline value' and puts a cross mark ('x') on

the monitor. Hence, if the FHR is 150 beats per minute, one should expect to see five crosses ('x') on the screen. This process of recording requires to be started when FHR is normal. If the computer finds that any recent information on fetal ECG after analysis is significantly different to its original calculation (i.e. the baseline T/QRS or ST segment values obtained in the initial four minutes), this will be flagged up as a 'ST Event' and appropriate action needs to be taken. Therefore, this type of monitoring cannot be initiated late or after a doubt in clinical settings STAN mainly analyses T/QRS ratios and 'ST segment changes' of fetal ECG complexes and produces two types of 'ST Events': 'T/QRS ST Events' and 'Biphasic ST Events'.

T/QRS ST events: It denotes periods of myocardial hypoxia that results in an increase in 'T Wave' height. If the hypoxic insult is short lasting, i.e less than 10 min, it is termed 'Episodic T/QRS Rise' and if the hypoxia is long lasting for more than 10 min, it is termed 'Baseline T/QRS rise'. This is probably due to a 'catecholamine surge' from the fetal adrenal gland that occurs secondary to hypoxic stress. Catecholamines cause tachycardia and also breakdown myocardial glycogen into glucose to increase the energy substrate availability for the myocardium to continue to function and supply adequate nutrients to the brain and adrenal gland of the fetus, both of which are essential for fetal survival. This process of 'catecholamine induce glycogenolysis' which causes release of potassium ions along with glucose into the myocardial cell results in 'hyperkalemia' producing 'tall T Waves' and an increase in T/QRS ratio.

Excessive fetal movements may also result in catecholamine surges which can cause 'Episodic T/QRS ST events'. In this case, the CTG would be otherwise normal and show accelerations and hence, these ST events can be neglected.

Biphasic ST events: ST Segment reflects a period of quiescence when the myocardium 'rests' just after a contraction (depolarisation), prior to relaxation (repolarisation). Under normal situations, myocardial cell membrane should not allow transfer of any ions during this 'absolute refractory period' and the ST segment would be 'iso-electric' and will have a stable baseline.

When there is a disturbance of myocardial function due to hypoxia, infection, structural heart defects, myocardial dystrophies or prematurity (less contractile elements), the ST Segment of the fetal ECG segment may shift upwards or downwards leading to 'Biphasic ST Events'. As the endocardium undergoes ischemia, sequence of repolarisation gets changed and the direction of current flow gets reversed. This results in depression of the ST segment of the fetal ECG complex with or without a negative T wave.

Breech presentation may also give rise to multiple biphasic ST events despite a normal healthy fetus as the heart is turned 'upside down' in relation to the external skin electrode that results in similar reversal in current flow in respect to the reference electrode. Some STAN Machines have a 'breech mode' to rectify this problem and this should be turned on if an assisted vaginal breech delivery is allowed and continuous electronic FHR monitoring is required.

STAN technology works by determining the 'normal baseline ECG' for the index fetus and then compares subsequent ECG complexes obtained from the fetus with the initial baseline. Hence, if the CTG is pre-terminal or if there is total loss of variability, the fetus may have already lost all the reserves to respond to hypoxia and there will be no further changes in the fetal ECG complexes that can be determined by STAN machine. This would mandate an immediate delivery to salvage the fetus.

In all other cases (normal CTG or CTG with decelerations and normal baseline heart rate and variability), the fetus can respond further to hypoxia (i.e. show further changes in ECG complexes) and hence, could be analysed by STAN Machine.

STAN are based on classification of both CTG and ST events. For this purpose CTG is classified as normal, intermediary, abnormal and preterminal CTG. This classification is very similar to the routine classification that is followed with only few differences.

According to these guidelines, a preterminal CTG irrespective of ST event requires an immediate intervention or delivery depending on the stage of labour.

In a patient with normal CTG despite any ST event, expectant management can be done.

In case of intermediary or abnormal CTG the ST events are observed and if there are any significant changes from baseline then intervention is required either in form of fetal scalp sampling or delivery. If the change is insignificant, then labour can be continued.[11]

Cochrane review 2015 showed that fetal ECG plus CTG resulted in fewer fetal blood sampling and surgical intervention when compared with CTG alone but there was no difference in the number of cesarean deliveries between the two groups. However, this highly sophisticated technology is still to become a part of routine clinical services.[11]

Points to Remember

- It is essential to know the history and the clinical information of the patient before interpreting a CTG for an accurate diagnosis and to avoid unnecessary interventions.
- If the tracing has a stable baseline and a reassuring variability, the risk of hypoxia to the fetus is very unlikely.
- The FHR graph should always be seen in conjunction with that of uterine contractions.
- In a stable patient with a normal tracing, one should interrupt the electronic fetal monitoring tracing for up to 30 minutes to facilitate ambulation, bathing or position change for the patient.
- In case of suspicious and pathological tracings intervention is required to avoid adverse neonatal outcome. An attempt must be made to identify the underlying cause for the appearance of the abnormal pattern and necessary measures can be initiated accordingly so that adequate fetal oxygenation is restored. The interventions always do not necessarily mean an immediate caes are an section or instrumental vaginal delivery. In case of abnormal tracings, digital fetal scalp stimulation can be performed and if positive acceleratory response is not seen, then FBS should be considered to determine the need for delivery.
- Fetal pulse oximetry and fetal ECG are newer monitoring techniques that are in research phase.
- Careful and objective interpretation of intrapartum CTG is essential to minimise unwarranted surgical intervention.

REFERENCES

1. Cousens S, Blencowe H, Stanton C, Chou D, Ahmed S, Steinhardt L, et al. National, regional, and worldwide estimates of stillbirth rates in 2009 with trends since 1995: a systematic analysis. Lancet 2011;377:1319–30.

2. Martis R, Emilia O, Nurdiati DS, Brown J. Intermittent auscultation (IA) of fetal heart rate in labour for fetal well-being. Cochrane Database Syst Rev. 2017 Feb 13;2:CD008680. doi: 10.1002/14651858.CD008680.pub2.

3. Blix E, Reiner LM, Klovning A, Oian P. Prognostic value of the labour admission test and its effectiveness compared with auscultation only: a systematic review. BJOG 2005;112(12):1595–604.

4. T Rowe, V Senikas, J Fairbanks. Fetal Health Surveillance: Antepartum and Intrapartum Consensus Guideline. JOGC 2007;29(9), Suppl 4.

5. Alfirevic Z, Devane D, Gyte GM, Cuthbert A. Continuous cardiotocography (CTG) as a form of electronic fetal monitoring (EFM) for fetal assessment during labour. Cochrane Database Syst Rev. 2017 Feb 3;2:CD006066. doi: 10.1002/14651858.CD006066.pub3.

6. Hadar E, T Biron-Shental, Gavish O, Raban O, Yogev Y. A comparison between electrical uterine monitor, tocodynamometer and intra uterine pressure catheter for uterine activity in labor. J Matern Fetal Neonatal Med 2014;28(12): 1367–74.

7. Martel MJ, MacKinnon CJ. Clinical practice obstetrics committee, society of obstetricians and gynaecologists of Canada. Guidelines for vaginal birth after previous caesarean birth. J Obstet Gynaecol Can 2005;27(2):164–88.

8. East CE, LeaderLR, Sheehan P, Henshall NE, Colditz PB, Lau R. Intrapartum fetal scalp lactate sampling for fetal assessment in the presence of a non-reassuring fetal heart rate trace. Cochrane Database Syst Rev 2015;(5):CD006174. doi: 10.1002/14651858.CD006174.pub3.

9. East CE, Begg L, Colditz PB, Lau R. Fetal pulse oximetry for fetal assessment in labour. Cochrane Database Syst Rev 2004;(4):CD004075.

10. Westerhuis MEMH, Visser GHA, Moons KGM, van Beek E, Benders MJ, Bijvoet SM, et al. Cardiotocography plus ST-analysis of fetal electrocardiogram compared with cardiotocography only for intrapartum monitoring: a randomized controlled trial. Obstet Gynecol 2010;115(6).

11. Neilson JP. Fetal electrocardiogram (ECG) for fetal monitoring during labour. Cochrane Database Syst Rev 2015;(12):CD000116. doi: 10.1002/14651858.CD000116.pub5.

Fetal Conditions and Related CTG Readings

Asmita M Rathore, Deepali Mittal Mishra

CTG has become a very important tool in monitoring of fetal condition and form base for intervention to prevent adverse fetal outcome. This chapter will cover cardiotocography readings associated with fetal conditions commonly encountered in clinical practice like fetal sleep, hypoxia, prematurity, cord compression, chorioamnionitis, anemia and fetal growth restriction. The CTG interpretation and description in this chapter is based on FIGO 2015 guidelines.

Fetal sleep: Fetal sleep is a physiological state associated with changes in CTG pattern. During deep fetal sleep, variability is usually in the lower range of normality, but the bandwidth amplitude is seldom under 5 bpm. It is important to look for 'cycling' periods of

normal variability and accelerations alternating with reduced baseline variability as depicted in Fig. 9.1. This denotes a fetus with a normal 'sleep-activity' pattern and indicates a neurologically intact, well oxygenated fetus. In such cases the reduced baseline variability lasts for about 30–50 minutes. Persistence of reduced variability for more than 50 minutes is considered abnormal.[1]

Fetal hypoxia: During labour a fetus is exposed to three types of hypoxia: Acute, subacute or gradually evolving hypoxia based on the onset and progression of hypoxic stress.

Acute hypoxia: Results in a sudden drop in baseline heart rate. It is termed 'single prolonged deceleration if it lasts for less than

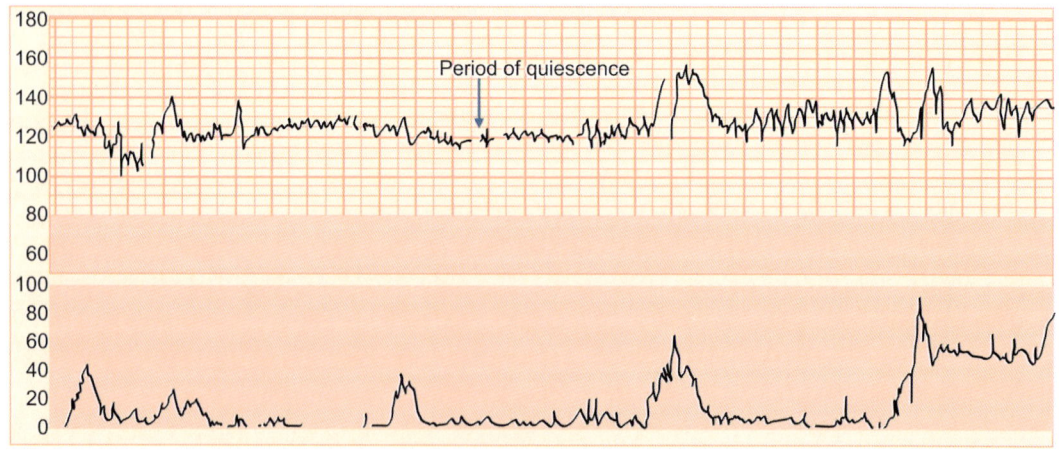

Fig. 9.1: CTG showing 'cycling': alternating periods of quiescence and activity.

3 minutes and then recovers to normal baseline. If the deceleration lasts for more than 3 minutes, it is termed 'prolonged decelerations lasting for more than 3 minutes'.[1] If the heart rate remains below 80 beats /minute for over 10 minutes, then it is termed prolonged baseline bradycardia. A depiction of acute hypoxia is shown in Fig. 9.2, showing sudden drop of fetal heart rate to 60 bpm with reduced variability. Clinicians should exclude three major 'accidents' during labour (abruption, cord prolapse and cesarean scar rupture) and two iatrogenic causes (uterine hyperstimulation due to oxytocin infusion and maternal hypotension usually secondary supine hypotension or epidural analgesia).[2] If there is any clinical evidence of these 'accidents', an immediate delivery should be undertaken to salvage the fetus. This is because metabolic acidosis is likely to get worse with time due to continued reduction in the uteroplacental circulation. In the presence of acute hypoxia, the fetal pH has been shown to drop at the rate of 0.01/minute.

Subacute hypoxia: In this situation, the fetus spends more time decelerating and progressively less time at the normal baseline FHR. Typically, the fetus spends less than 30 seconds at the baseline to 'wash off' carbon dioxide and acid and spends over 90 seconds building up carbon dioxide and acid (as shown in Fig. 9.3). pH of the fetus has been shown to drop at the rate of 0.01 every 2–3 minutes.[3] This situation is usually seen during active maternal pushing or with use of oxytocin to augment labour.[2]

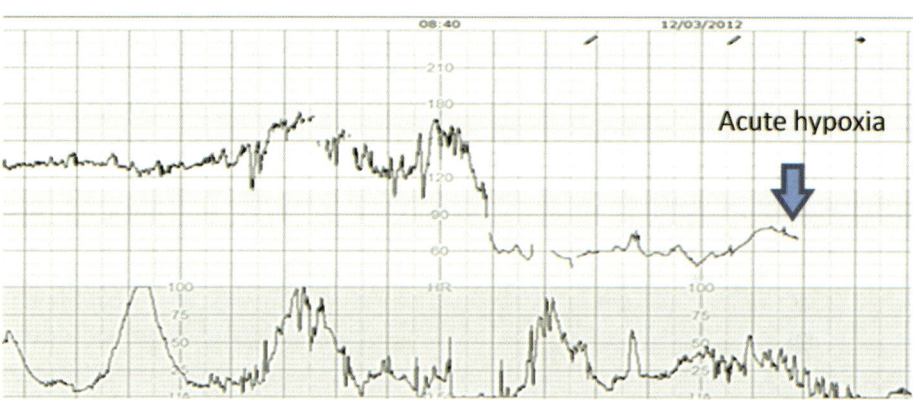

Fig. 9.2: Acute Hypoxia: Sudden bradycardia with loss of baseline variability.

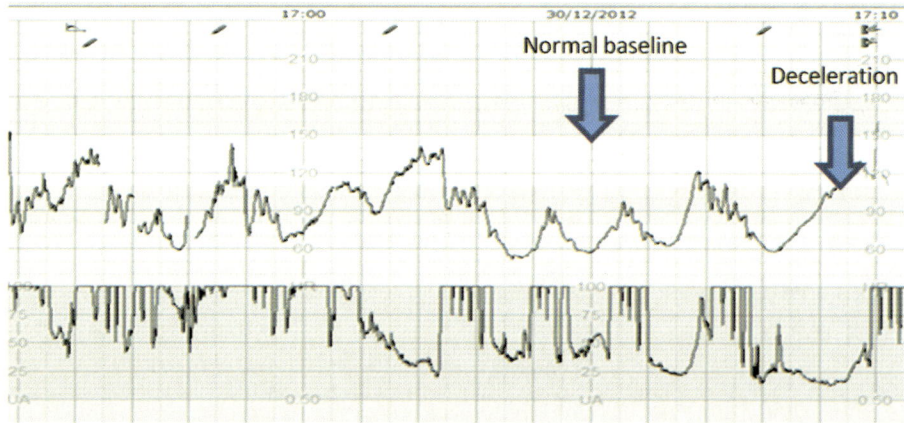

Fig. 9.3: Subacute Hypoxia: Progressively less time at the normal baseline (<30 s) as compared with decelerations.

Gradually evolving hypoxia: Hypoxic stress may develop over hours thus giving the fetus an opportunity to utilize its compensatory mechanisms to avoid hypoxic injury. In this scenario, CTG would initially show decelerations followed by disappearance of accelerations as the fetus attempts to conserve energy by limiting muscle activity that may increase oxygen requirement. If the hypoxic insult continues (in the form of spontaneous contraction on oxytocin induced), fetus then releases catecholamines to increase the heart rate and its cardiac output to supply vital organs. Figure 9.4 depicts the CTG changes as seen in gradually evolving hypoxia showing increase in baseline FHR with repetitive deceleration. Despite of fetal efforts at compensation, if the hypoxic insult persists, then decompensation ensues resulting in reduced perfusion of brain leading to loss of baseline variability. Finally, lack of oxygenation of the coronary arteries that arise at the root of the ascending aorta would lead to myocardial hypoxia and acidosis leading to a 'step-ladder pattern to death'. After such repeated attempts to return to the baseline, a terminal bradycardia will ensue leading to fetal death.[2]

Chronic hypoxia: In this situation, a fetus has been exposed to a prolonged period of hypoxia during the antenatal period usually secondary to a chronic uteroplacental insufficiency. Usual intrauterine adaptations include reduction in growth, movements and diversion of oxygenated blood and nutrients from nonvital organs to supply the vital organs. As the hypoxic insult has occurred at some point during the antenatal period (i.e. prior to the onset of labour) the CTG often shows a higher baseline with reduced variability and shallow decelerations with uterine contractions (as shown in Fig. 9.5). The presence of this CTG pattern requires immediate delivery because

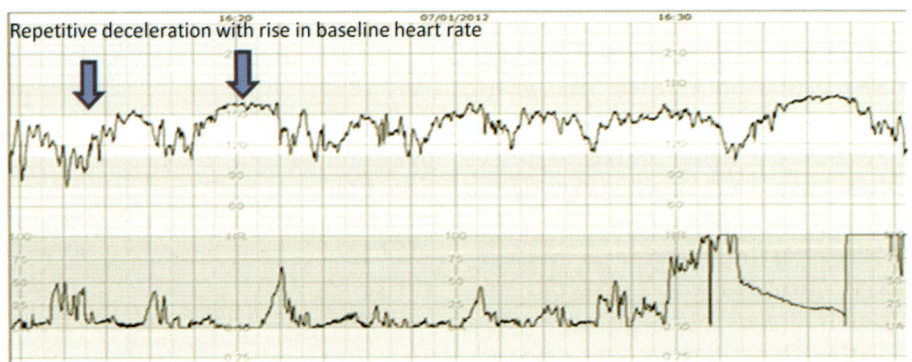

Fig. 9.4: Gradually evolving hypoxia: Repetitive deceleration with absence of accelerations and a rise in the baseline fetal heart rate (due to catecholamine surge).

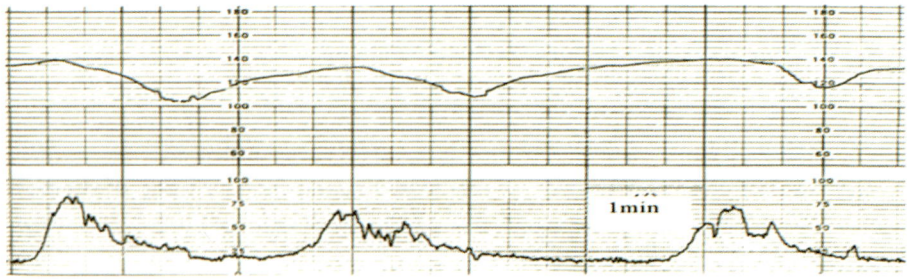

Fig. 9.5: Chronic hypoxia: Absence of acceleration and presence of late deceleration with reduced variability.

with the onset of uterine contractions, and resultant intermittent umbilical cord compression and reduction in uteroplacental circulation, there will be a further reduction in oxygenation leading to HIE as well as myocardial failure leading to a terminal bradycardia.[3]

Uterine hyperstimulation: Uterine hyperstimulation refers to any increase in uterine activity (frequency, strength and duration) that is associated with changes in the FHR on the CTG Trace. This is secondary to both repeated and prolonged umbilical cord compression as well as sustained reduction of oxygen supply to the placental venous sinuses secondary to increase in the duration of contraction and decreased relaxation time for replenishment of oxygen. Fetal decompensation can rapidly ensue during uterine hyperstimulation to rapid and progressive reduction in fetal oxygenation. Fetuses respond by reducing myocardial workload (decelerations) to conserve oxygen and demonstrate variable decelerations (umbilical cord compression) and late decelerations (lack of oxygen within the placental venous sinuses) and attempt to increase the heart rate by releasing adrenaline in order to continue perfusing the central organs (brain and the heart muscle). Figure 9.6 shows an increase in baseline FHR with uterine hyperstimulation as a compensatory mechanism to progressive fetal hypoxia. If these compensatory mechanisms

fail, a reduction of baseline variability (loss of blood supply to brain centres) and a gradual and progressive reduction in baseline FHR (lack of oxygen to the heart muscle which is unable to keep pumping blood at the same rate) occurs. If timely and appropriate interventions to reduce hypoxic stress and/or to expedite delivery are not instituted, then myocardial failure occurs (terminal bradycardia).[3]

Preterm Fetus

Characteristics of antepartum and intrapartum FHR tracings differ in the preterm fetus as compared to a term fetus. Notably, fetal baseline heart rate is higher, averaging at 155 between 20 and 24 weeks (compared to a term fetus where average baseline FHR is 140). Figure 9.7 show CTG changes of preterm fetus with baseline of 160 bpm and low amplitude acceleration. With advancing gestational age, there is a gradual decrease in baseline FHR. These findings are likely to reflect fetal immaturity, as the basal heart rate is the result of counteraction between parasympathetic, and sympathetic systems. As the fetus develops beyond 30 weeks, the progressive increase in the parasympathetic influence on FHR results in a gradual lowering of baseline rate.[4]

Fetal heart rate accelerations are also noted to change with advancing gestational age. Accelerations of FHR in association with fetal movements occur as a result of fetal somatic

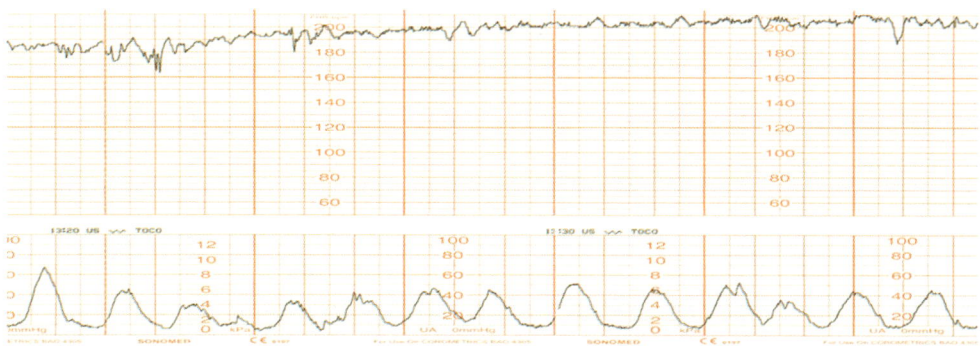

Fig. 9.6: Uterine hyperstimulation: Increased frequency of uterine contractions with rise in fetal baseline heart rate and decelerations.

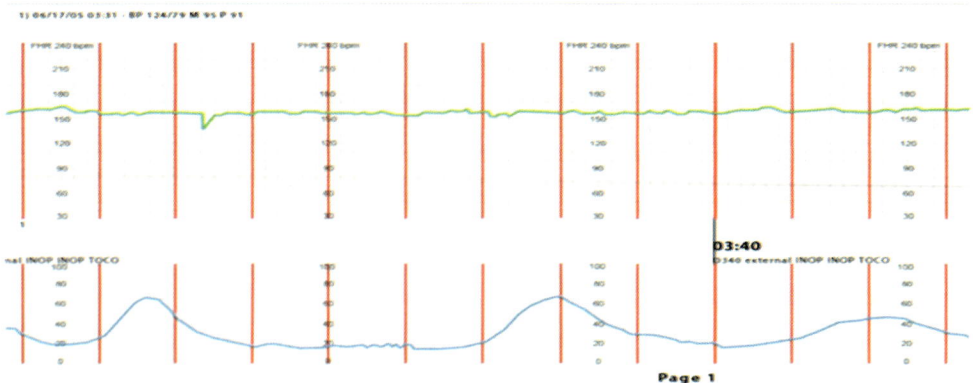

1) 06/17/05 03:31 · BP 124/79 M 95 P 91

03:40

0340 external INOP INOP TOCO

Page 1

Fig. 9.7: Baseline tachycardia with low amplitude acceleration.

activity and are first apparent in the 2nd trimester. Before 30 weeks of gestational age, the frequency and amplitude of accelerations are reduced. Preterm fetus may exhibit accelerations with a peak of only 10 beats per minute lasting for 10 seconds.[5] With subsequent increase in gestational age, the frequency of accelerations increases along with amplitude over the baseline value.[5]

Fetal heart rate decelerations in the absence of uterine contractions often occur in the normal preterm fetus between 20 and 30 weeks gestation. These decelerations have a lower depth and duration, but can be seen frequently on intrapartum CTG tracings. Variable decelerations have been shown to occur in 70–75% of intrapartum preterm patients, in comparison to the term patient where an intrapartum rate of 30–50% is seen.[6] Several theories have been proposed as a potential explanation for this FHR pattern, notably decreased amount of amniotic fluid, reduced Wharton jelly component in the cord of the preterm fetus and lack of development of the fetal myocardium and, therefore, the resultant reduced force of contraction.

Baseline variability may be affected due to incomplete development of autonomic nervous system and subsequent interplay between parasympathetic and sympathetic systems. Variability may also be decreased secondary to the effect of fetal tachycardia

present in preterm fetuses. Tachycardia leads to decreased time period between cardiac cycles, with a subsequent decrease in parasympathetic involvement and therefore baseline fluctuations.

One of the hallmarks of fetal well-being is considered to be "cycling" of the FHR.[1] As the maturity of the central nervous system occurs with advancing gestational age, this "cycling" of the FHR is established. Hence, in an extreme preterm infant, cycling may be absent and this may be due to functional immaturity of the central nervous system, rather than hypoxic insult.

With increasing gestation the baseline FHR is likely to decrease from the upper limits of the normal range. Baseline variability of greater than five beats per minute with signs of cycling is likely to develop, between 30 and 32 weeks gestation. The predominance of variable decelerative patterns should initially reduce and disappear after 30 weeks gestation. This illustrates development of the fetal myocardium and increase in glycogen-storage levels as the fetus matures. Persistence of late decelerations within this cohort is likely to represent ongoing uteroplacental insufficiency. In this situation, the blood flow within the intervillous space is decreased resulting in accumulation of carbon dioxide and hydrogen ion concentrations. In the non-compromised, non-acidic fetus, intermittent hypoxia results in decelerations with subsequent transient fetal hypertension. With passage of time, continuation of this hypoxic insult will lead to acidemia, loss of initial "compensatory" hypertensive response, and may proceed to cause permanent cerebral injury. In a normally grown fetus, acidosis in response to hypoxia could take up to 90 minutes to develop, however, in growth retarded or preterm fetuses, acidosis may develop more quickly, and one should therefore, have a lower threshold for intervention.

Cord Compression

The most common pattern seen on CTG with cord compression is variable decelerations. Variable decelerations (as shown in Fig. 9.8) are so named because they vary in shape, form and timing in relation to the uterine contractions. Typical variable decelerations consist of a slight rise in the FHR (called shouldering) both before and after the deceleration. Simple variable deceleration last for less than 60 seconds in the presence of normal baseline heart rate and variability. Atypical variable deceleration last for more than 60 seconds and may lose their shouldering, have a slow recovery, have an overshoot, or be combined with a late deceleration. Repeated variable deceleration may indicate oligohydramnios.[1]

Fetal Anemia

Severe fetal anemia is indicated by sinusoidal pattern on CTG, as occurs in cases of anti-D allo-immunization, fetal-maternal hemorrhage, twin-to-twin transfusion syndrome and ruptured vasa previa.[1] The sinusoidal pattern is rare but ominous and is associated with high rates of fetal morbidity and mortality. The sinusoidal pattern (shown in Fig. 9.9) is a regular, smooth, undulating form typical of a sine wave that occurs with frequency of two to five cycles per minute and an amplitude range of 5 to 15 bpm. It is also characterized by a stable baseline heart rate of 120 to 160 bpm and absent beat-to-beat variability. It should be differentiated from the "pseudosinusoidal" pattern, which is a benign, uniform long-term variability pattern. A pseudosinusoidal pattern shows less regularity in the shape and amplitude of the variability waves and the presence of beat-to-beat variability, compared with the true sinusoidal pattern.

Fetal Growth Restriction

A derangement in trophoblastic differentiation is thought to underlie the pathophysiology of gestational hypertension, pre-eclampsia, and Fetal Growth Restriction (FGR).

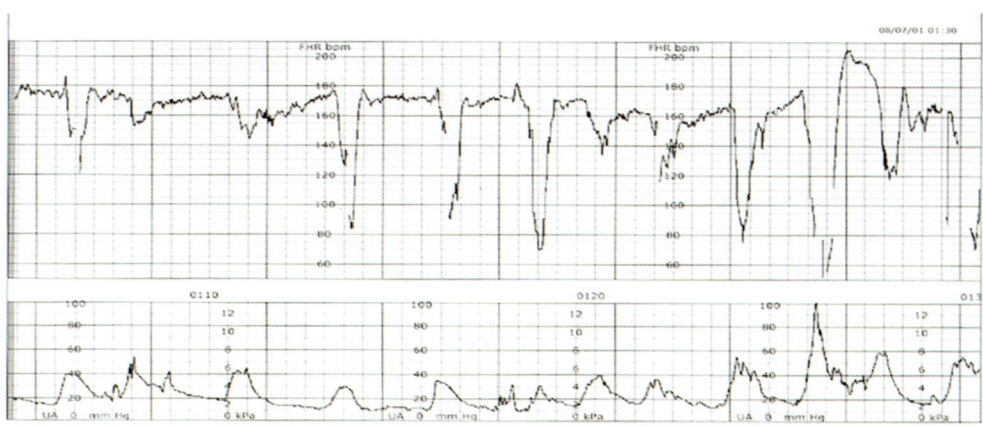

Fig. 9.8: Variable deceleration: Decelerations with normal variability and no association with uterine contractions.

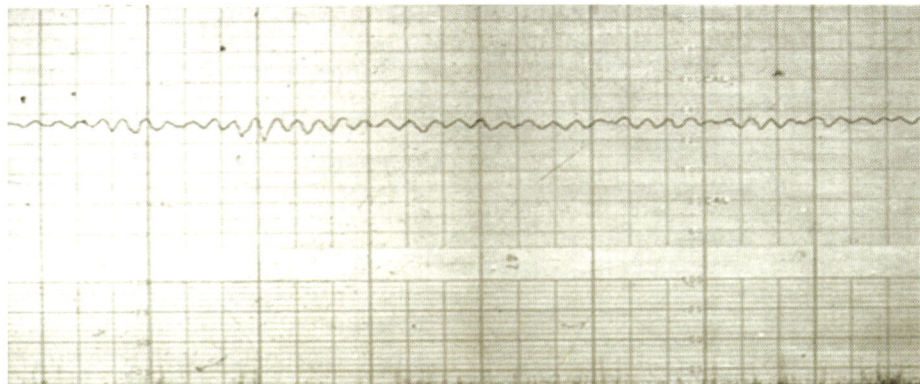

Fig. 9.9: Sinusoidal baseline.

Defective implantation leads to uteroplacental insufficiency. Uteroplacental insufficiency is indicated by late decelerations (as depicted in Fig. 9.5), mediated through the fetal chemo-receptor mechanism secondary to fetal hypoxemia. Oxygenated blood needs to reach the uteroplacental venous sinuses to remove the stimulus of hypoxia to the chemo-receptors. Hence, there is a delay in recovery of the FHR to the baseline rate. These are characterized by late recovery; the nadir of these decelerations is seen after the peak of uterine contractions and they reach the baseline at least 20 seconds after the contrac-tion wanes.[3] Due to its delayed recovery to the baseline rate, it is called 'late'. It is impor-tant to exclude fetal acidosis in the presence of late decelerations if labour is allowed to continue.

Chorioamnionitis

Chorioamnionitis can lead to an increase in baseline FHR coupled with decreased vari-ability, even before the rise in maternal temperature is recorded. Particularly in pre-term infants, persistent fetal tachycardia over a longer period of time triggered by endo-toxins should prompt attendant gynecologists to consider early delivery of the infant as there is a known correlation between fetal brain damage and chorioamnionitis.[1]

Postdated Pregnancy

In post-term fetus baseline bradycardia is seen, i.e. a decrease in FHR below 110 beats/ minute. It may be a physiological finding in post-term fetus, and the fetus is unlikely to be hypoxic provided that there are acceleration and normal baseline variability. In the absence of acceleration and normal baseline variability, it may be a sign of fetal hypoxia.If bradycardia last for less than 3 minutes it is considered suspicious, and if it is for more than 3 minutes it is considered to be abnormal and is an indication to consider remedial actions or delivery.[1]

Transplacental Drugs

Many medications easily pass the placental barrier to reach the brain and other centers of circulatory regulation at high concentrations.

- Sedatives, anesthetics (both general and local) and antiepileptic drugs reduce heart rate variability and result in flatter curves, this occurs with corticosteroids (dexame-thasone and betamethasone) and cocaine abuse. Magnesium sulfate has also been associated with reduced FHR variability.[1]
- Beta mimetics (e.g. fenoterol, salbutamol) which are used for tocolysis are mostly metabolized prior to the placental barrier, but they can still be effective in minute

quantities or as metabolites, leading to an increase in FHR with a simultaneous reduction of variability and heart rate accelerations. Such CTG patterns are usually reversible after 5–7 days at the latest and do not constitute a concrete fetal risk.[1]

- Antihypertensives such as beta blockers cross the placental barrier on a 1:1 basis and, depending on the dose, can result in complete blockage of the fetal sympathetic nervous system. This leads to flattening of accelerations with pronounced bradycardia or even tachycardia. Beta blockade can also impair fetal circulatory centralization and glucose mobilization, which are important if there is a lack of oxygen.[1]
- Epidural analgesia may also cause a rise in maternal temperature resulting in fetal tachycardia.[1]

Fetal Cardiac Abnormalities

Fetal cardiac abnormalities can present with varied presentation on CTG depending on type of cardiac lesion and arrhythmia. Fetal arrhythmias such as supraventricular tachycardia and atrial flutter present as tachycardia on CTG while in atrial-ventricular block bradycardia is seen. Fetus with cardiac anomalies usually shows decreased beat to beat variability. Fetus with cardiac malformations can show sinusoidal pattern on CTG.[1]

REFERENCES

1. Visser GH, Ayres-de-Campos D. FIGO Intrapartum Fetal Monitoring Expert Consensus Panel. FIGO consensus guidelines on intrapartum fetal monitoring: Adjunctive technologies. Int J Gynaecol Obstet 2015;131(1):25–9.
2. Pinas A, Chandraharan E. Continuous cardiotocography during labour: Analysis, classification and management. Best Pract Res Clin Obstet Gynaecol 2016;30:33–47.
3. Chandraharan E, Arulkumaran S. Prevention of birth asphyxia: responding appropriately to cardiotocograph (CTG) traces. Best Pract Res Clin Obstet Gynaecol 2007;21(4):609–24.
4. Afors K, Chandraharan E. Use of Continuous Electronic Fetal Monitoring in a Preterm Fetus: clinical dilemmas and recommendations for practice. J Pregnancy 2011;Article ID 848794.
5. Wheeler T, Murrills A. "Patterns of fetal heart rate during normal pregnancy," British Journal of Obstetrics and Gynaecology 1978;85(1):18–27.
6. Zanini B, Paul RH, Huey JR. "Intrapartum fetal heart rate: correlation with scalp pH in the preterm fetus," American Journal of Obstetrics and Gynecology 1980;143(1):952–7.

Analysis of the Cardiotocography Trace

Gita Arjun

In the 1960s, continuous cardiotocography (CTG) as a form of electronic fetal monitoring (EFM) was introduced for obtaining more accurate information of fetal status, and the early diagnosis of fetal hypoxia. An abnormal tracing prompted intervention when required.

Normal human labour typically consists of regular uterine contractions. During each contraction, there are repeated, though transient, episodes of decreased fetal oxygenation. Most fetuses tolerate this process well since they are well-oxygenated. Fetuses with compromised placental function may not tolerate the transient drops in oxygenation during contractions. The fetal heart rate (FHR) pattern recorded during cardiotocography helps to identify the fetus at risk. A CTG trace is 'an indirect marker of fetal cardiac and central nervous system responses to changes in fetal blood pressure, blood gases, and acid-base status'.[1]

This chapter will deal with the analysis and interpretation of both intrapartum and antepartum FHR monitoring.

INTRAPARTUM FETAL HEART RATE TRACE

Rationale for Intrapartum FHR Monitoring

The rationale for intrapartum FHR monitoring is that identification of FHR patterns potentially associated with inadequate fetal oxygenation may enable timely intervention

that would reduce the likelihood of hypoxic injury or death.

Although virtually all published obstetric guidelines advise monitoring the FHR during labour, the benefit of this intervention has not been clearly demonstrated and this position is largely based upon expert opinion and medicolegal precedent. A 2017 systematic review of CTG during labour[2] stated that it is associated with reduced rates of neonatal seizures, but no clear differences in cerebral palsy, infant mortality or other standard measures of neonatal well-being have been demonstrated. On the other hand, the review showed that EFM is associated with an increase in interventions, including cesarean section, vaginal operative delivery, and the use of anesthesia. No difference in the long-term outcome has been demonstrated.

Cardiotocography is useful in assessing FHR patterns when: (1) Intermittent Auscultation (IA) picks up abnormalities or (2) there is meconium present in the amniotic fluid. In these situations, a reassuring FHR tracing allows the obstetrician to continue monitoring labour without any major intervention.

It is reasonable clinical practice to use intermittent auscultation during labour to pick up FHR abnormalities. When FHR abnormalities occur, CTG is initiated. Continuous monitoring with CTG has not been shown to improve perinatal outcomes in low-risk women.

The Fetus at-risk for FHR Abnormalities

Typically, a well-oxygenated term fetus usually tolerates the transient decrease in oxygenation during contractions, and shows no adverse effect. However, even a well-compensated, term fetus may respond poorly to a prolonged period of decreased oxygenation as may occur in abruption, bleeding due to placenta previa, supine hypotension, and hypotension associated with epidural analgesia.

In conditions with chronic placental insufficiency such as hypertension, diabetes, and the antiphospholipid antibody syndrome where there is fetal growth restriction, the fetus may not tolerate the decrease in oxygenation and may show signs of hypoxia. The preterm fetus too tolerates hypoxia poorly.

Analysis of CTG Trace

Terminology

Fetal heart rate patterns are described using the following terms:
- Baseline heart rate
- Baseline variability
- Periodic changes
  - Decelerations
  - Accelerations.

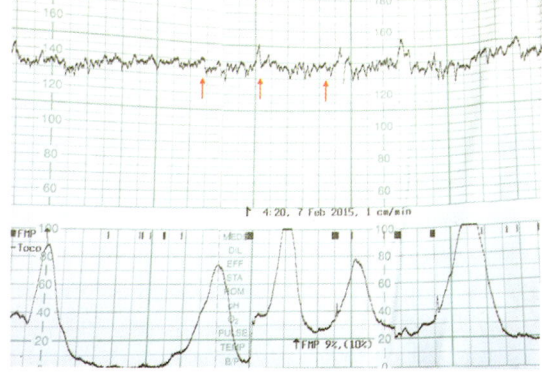

Fig. 10.1: Baseline heart rate on cardiotocograph: Fetal heart rate tracing showing a baseline heart rate of 130 bpm. Baseline variability is normal (5–25 bpm) and denoted by arrows.

Baseline heart rate: The normal baseline FHR is 110–160 beats/minute (bpm). Within this range, the baseline heart rate denotes the absence of any pathological factors that could influence FHR. Figure 10.1 shows the baseline FHR and baseline variability.

When the FHR exceeds 160 bpm for 10 minutes or more, it is called tachycardia. When the FHR is below 110 bpm for 10 minutes or more, it is called bradycardia (Box 10.1).

Box 10.1: Characteristics of tachycardia and bradycardia

- Tachycardia—Heart rate of >160 beats/minute.
  - Causes of tachycardia
    - Maternal fever
    - Chorioamnionitis
    - Beta sympathomimetics
    - Fetal compromise
- Bradycardia—Heart rate of <110 beats/minute
  - Causes of bradycardia
    - Head compression
    - Fetal compromise

Baseline variability refers to the fluctuations in the baseline FHR that are irregular in amplitude and frequency. The normal baseline variability has an amplitude range of 5–25 bpm. It is measured from peak to trough. Fetal heart rate variability is a reflection of modulation of heart rate by the Central Nervous System (CNS) and the autonomic nervous system (Box 10.2).

Box 10.2: Classification of baseline variability

Minimal baseline variability	<5 beats/minute
Moderate or normal	5–25 beats/minute
Marked	>25 beats/minute

Decreased variability is an important sign of fetal hypoxia. Decreased or absent baseline variability occurs due to:
- Maternal administration of
  - Analgesics
  - Sedatives
  - Magnesium sulfate
- Fetal hypoxia.

Periodic Changes

Accelerations: Accelerations are transient increases in basal heart rate by >15 bpm, lasting for at least 15 seconds. They are a reassuring sign (Fig. 10.2), and the presence of accelerations rules out fetal hypoxia.[3] However, it is important to remember that the absence of accelerations does not necessarily denote fetal hypoxia.

Decelerations: CTG can differentiate decelerations into three types based on their relationship to uterine contractions, and two that may or may not be related to uterine contractions. A deceleration is considered significant if it decreases >15 bpm below baseline, lasts for >15 seconds, and is repetitive.

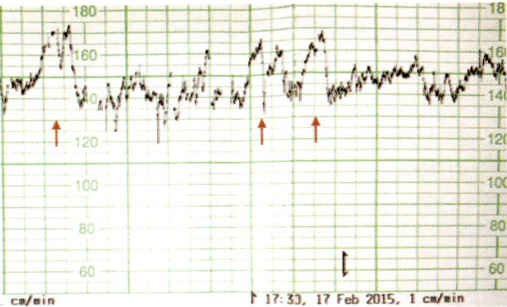

Fig. 10.2: Accelerations (transient increases in baseline heart rate) are marked by arrows.

- **Early decelerations:** These decelerations are benign occurrences and are unrelated to fetal hypoxia. They are symmetric gradual drops in the FHR that mirror the uterine contraction. The nadir (the lowest point) of the deceleration coincides with the peak of the contraction. They represent an autonomic response to head compression though the exact mechanism is not known (Fig. 10.3).

- **Late decelerations:** These decelerations are caused by placental insufficiency leading to fetal hypoxia.[4] They commence after the start of the contraction and return to the baseline after the contraction is over. The nadir of the deceleration occurs after the peak of the contraction (Fig. 10.4).

- **Variable decelerations:** They are characteristically variable in duration, intensity, and timing. They represent fetal autonomic reflex response to transient mechanical compression of the umbilical cord.[5] They resemble the letter 'U, ''V,' or 'W' and may be variable even in relation to the uterine contraction (Fig. 10.5). Intermittent variable decelerations are often seen even in a normal labour tracing, but the fetus tolerates transient cord compression. However, persistent, deep, and recurrent variable

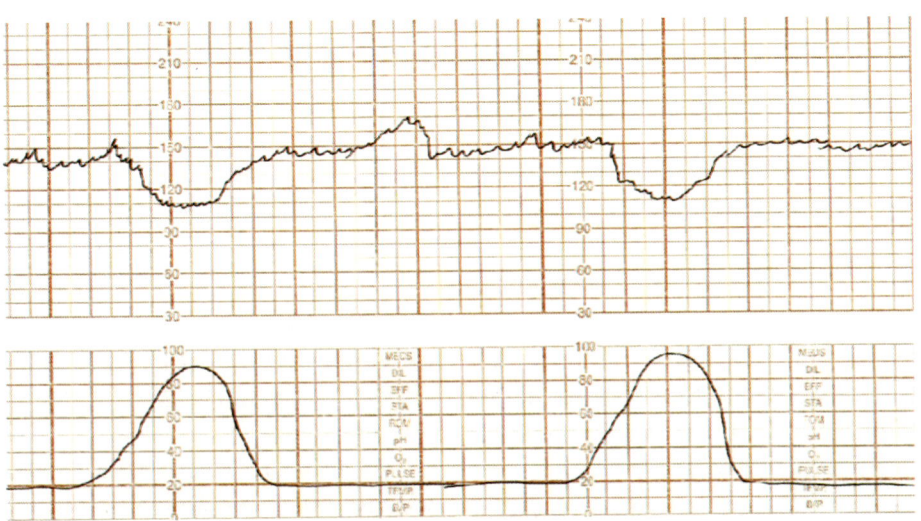

Fig. 10.3: Early decelerations. The decelerations are symmetric and mirror the contraction.

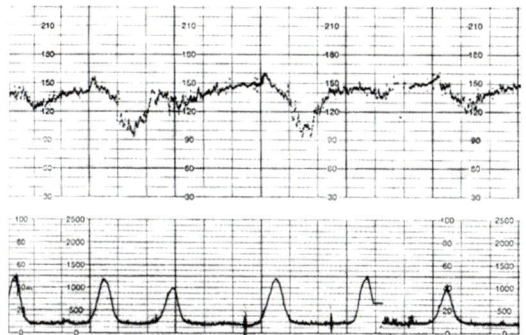

Fig. 10.4: Late decelerations. The decelerations are asymmetric and return to baseline after the contraction is over.

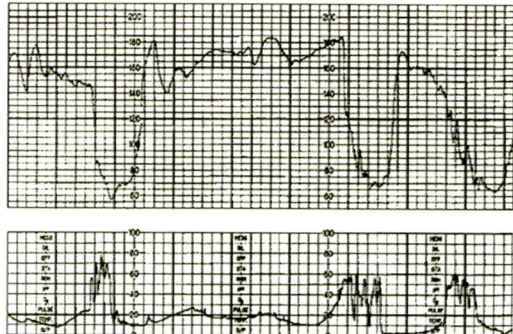

Fig. 10.5: Variable decelerations. Repetitive variable decelerations are seen on this tracing.

decelerations are indicative of fetal acidosis. They vary in onset, depth, and duration.

The following types of decelerations may also occur:

- **Prolonged deceleration:** Prolonged deceleration is one where the deceleration lasts > 2 minutes but < 10 minutes. If it continues for > 10 minutes, it is considered a shift of the baseline heart rate (bradycardia). It is indicative of prolonged cord compression, hypotension, or severe, acute placental insufficiency.[6]

- **Sinusoidal pattern:** This is a smooth, sine wave-like undulating pattern in the baseline FHR with a cycle frequency of 3–5 per minute that persists for 20 minutes or more. The presence of a sinusoidal pattern is indicative of fetal anemia or severe hypoxia/acidosis.[7] The characteristic features of sinusoidal pattern are summarized in Box 10.3.

Box 10.3: Features of sinusoidal pattern

- Baseline heart rate of 120–160 bpm
- Markedly decreased or absent variability
- Oscillations varying between 5 and 15 bpm
- Cycle frequency of 3–5 times/min
- No accelerations seen

The analysis of the intrapartum CTG trace is summarized in Box 10.4.

Box 10.4: Analysis of the cardiotocograph trace

Analysis of the cardiotocograph trace requires mention of the following:

- Baseline FHR (in bpm)
- Baseline variability (normal, decreased, absent)
- Presence of accelerations (duration and elevation above baseline)
- Presence of decelerations (duration, decrease below baseline, and relation to contraction)
  - Early
  - Late
  - Variable

Three-tiered Fetal Heart Rate Interpretation System

Electronic fetal monitoring tracings are classified as:

- Reassuring (Category I by ACOG or normal by RCOG),
- Non-reassuring (Category III by ACOG or pathological by RCOG),
- Indeterminate (Category II by ACOG or suspicious by RCOG)

Category I or reassuring tracings are considered 'normal' since they are associated with a normal fetus which is tolerating labour well. There is no fetal acidemia. A Category I tracing will have all of the following:

- A baseline FHR of 110–160 bpm
- Absence of late or variable FHR decelerations
- Moderate FHR variability (6–25 bpm)
- FHR accelerations appropriate to gestational age.

Category III or non-reassuring tracings are considered 'abnormal' because they are associated with an increased risk of fetal hypoxic acidemia, which can lead to cerebral palsy and neonatal hypoxic ischemic encephalopathy. Findings on the tracings include:

- Absent baseline FHR variability and any of the following:
  - Recurrent late deceleration
  - Recurrent variable decelerations
  - Bradycardia
- Sinusoidal pattern.

Category II tracings or indeterminate FHR patterns neither suggest acidosis nor give clear indication of fetal well-being. They occur at some point in almost 84% of tracings.[8] They include the following:

- Tachycardia
- Minimal or marked variability
- Absent variability without recurrent decelerations
- Absence of accelerations without absent variability
- Recurrent late or variable decelerations without absent variability.

Evaluation and management of abnormal tracings is discussed in Chapter 11, Abnormal CTG—when to deliver.

ANTEPARTUM FETAL HEART RATE TRACE

The aim of antepartum fetal surveillance is to prevent fetal demise. In the presence of risk factors that may affect fetal well-being, antepartum fetal surveillance gives information about the intrauterine status of the fetus. Fetal heart rate assessment using tocography is an important part of antepartum fetal surveillance.

Antepartum fetal surveillance techniques are useful in assessing the risk of fetal death in pregnancies complicated by pre-existing maternal conditions (e.g. hypertension) as well as those in which complications have developed (e.g. fetal growth restriction). When an antepartum fetal test is abnormal, it is called a nonreassuring test.

Nonstress Test

The Nonstress Test (NST) is the most commonly performed method of antepartum fetal assessment. The FHR is recorded in the absence of contractions. Accelerations of the FHR are looked for.

The optimal frequency for performing the NST is not clearly defined[9] and may vary from daily to weekly, depending on the level of fetal compromise.

Rationale for Performing the Nonstress Test

The heart rate of the healthy fetus will temporarily accelerate with fetal movement. This will usually start after 26–28 gestational weeks, when the fetus begins to have neurological maturity.[1]

A fetus that is acidotic or neurologically depressed will not show accelerations on CTG. A reactive NST will demonstrate accelerations with fetal movements. Heart rate reactivity is considered a good indicator of normal fetal autonomic function. Loss of reactivity is associated most commonly with a fetal sleep cycle but may result from any cause of central nervous system depression, including fetal acidosis (Box 10.5).

> **Box 10.5: Fetal heart rate accelerations in the nonstress test**
>
> - Presence
>   - Well-oxygenated fetus
>   - 15 bpm above baseline
>   - Lasting for at least 15 seconds
> - Absence
>   - Fetal sleep
>   - Narcotic medications to the mother
>   - Fetal hypoxemia
>   - Fetal acidosis

Interpretation

An NST is interpreted as follows:

- **Reactive** (Fig. 10.6): From 32 weeks to term, an NST is defined as reactive if two or more FHR accelerations occur, reaching a peak of

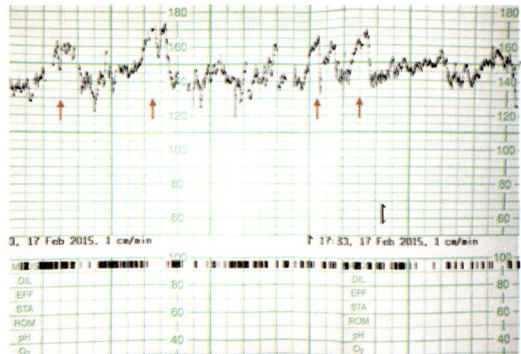

Fig. 10.6: Reactive non-stress test showing accelerations. The waveform represents the fetal heart rate. The arrows point to accelerations.

at least 15 beats per minute (bpm) above the baseline rate and lasting for at least 15 seconds from onset to return in a 20 minute period.[10] This is summarized in Box 10.6. The duration of the test should be extended to 40 minutes if there are no accelerations.

Box 10.6: Reactive non-stress test
- 2 or more acceleration
  - Recorded over 20 minutes
  - Up to 40 minutes if no accelerations
- Peak at least 15 bpm above baseline
- Lasting at least 15 seconds

- **Nonreactive** (Fig. 10.7): A nonreactive NST is defined as one that does not show accelerations over a 40 minute period. A nonreactive test may indicate fetal hypoxemia or acidosis. Additional tests such as vibroacoustic stimulation or a biophysical profile (BPP) may be needed to confirm that the fetal condition is nonreassuring (Box 10.7).

Box 10.7: Nonreactive nonstress test
- No accelerations in 40 minutes
- May be repeated after feeding the mother
- Vibroacoustic or BPP required to confirm
- Decelerations lasting 1 minute: significant

- **Decelerations during an NST:** FHR decelerations during an NST that persist for 1 minute or longer are significant and are associated with an increased risk of both cesarean delivery and fetal demise, especially when associated with oligohydramnios.[11]

Significance of decelerations in the presence of reactive NST: The majority of FHR decelerations during NST are variable decelerations, due to transient episodes of umbilical cord compression. The significance of decelerations in the presence of a reactive NST is uncertain. Several observational studies have noted an increased frequency of intrapartum FHR decelerations and operative delivery when this combination occurs.[10,11] In the majority, outcomes are good.

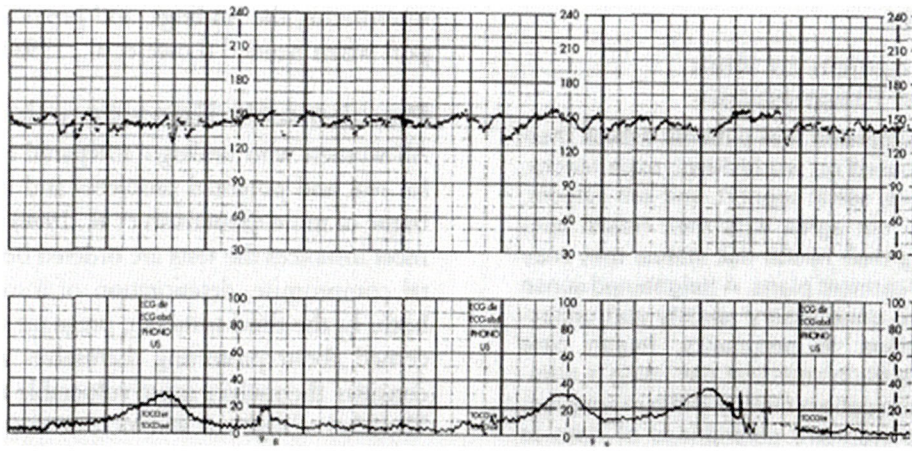

Fig. 10.7: Nonreactive non-stress test showing no accelerations. The upper waveform represents the fetal heart rate.

Prolonged decelerations, late decelerations, or variable decelerations during an NST require further evaluation, usually with ultrasound and Doppler evaluation. These additional tests will guide management.

A nonreactive NST may be due to fetal sleep or prematurity but may be indicative of fetal hypoxia or sepsis (Box 10.8). Hence, a nonreactive NST needs further evaluation.

Box 10.8: Causes for nonreactive nonstress test

- Associated with:
  - Fetal immaturity
  - Quiet fetal sleep
  - Fetal hypoxemia or acidosis
  - Fetal neurological or cardiac anomalies
  - Fetal sepsis
  - Maternal ingestion of drugs with cardiac effects

Other findings on NST

Bradycardia: Significant bradycardia is associated with increased perinatal mortality and morbidity and has a higher positive predictive value than a nonreactive NST. In the presence of bradycardia, further evaluation by a biophysical profile may be indicated.

Tachycardia: Preterm fetuses have a higher baseline FHR. In a term fetus, tachycardia may be due to maternal fever, fetal hypoxemia or acidosis.

Loss of variability: Loss of variability may occur with maternal sedation. When loss of variability occurs along with baseline tachycardia, it is indicative of fetal acidosis and requires further evaluation.

Predictive value of NST: The following points should be noted regarding the predictive value of an NST:

- A NST predicts the fetal status for the next 72 hours; therefore, in high-risk pregnancies such as postmaturity, diabetes mellitus, or severe hypertension, the test should be performed twice a week.
- A reactive NST has a higher predictive value. The false-negative rate is approximately 0.2–0.8%.

- A nonreactive NST has a false-positive rate of 50–60%.[1] This means that half the fetuses showing a nonreactive pattern may actually be well oxygenated. Hence, fetuses with a nonreactive NST should be evaluated further and management decisions should not be based on the NST alone.
- A NST cannot predict sudden events such as placental abruption or cord accidents.

Contraction Stress Test

The Contraction Stress Test (CST) is performed by recording the FHR in the presence of induced contractions. The response of the FHR is noted in relation to the contractions.

Rationale for CST: During a uterine contraction, there is a transient decrease in fetal oxygenation. If a fetus is already hypoxemic, the intermittent worsening in oxygenation during a uterine contraction will result in late decelerations of the FHR. Uterine contractions may also cause fetal umbilical cord compression in the presence of decreased amniotic fluid (oligohydramnios) in a high-risk pregnancy. This can result in variable decelerations.

The criteria for interpretation of a CST are given in Box 10.9.

Box 10.9: Interpretation of contraction stress test

- Positive (nonreassuring)
  - Late decelerations following 50% or more contractions
- **Negative (reassuring)**
  - No late or significant variable decelerations
- **Equivocal-suspicious**
  - Intermittent late decelerations or,
  - Significant variable decelerations
- **Equivocal-hyperstimulatory**
  - Decelerations that occur in the presence of hyperstimulation
    - Contractions >6 in 10 minutes,
    - Contractions lasting >90 seconds
- **Unsatisfactory**
  - Tracing is uninterpretable or,
  - Contractions <3 in 10 minutes

The analysis of FHR tracings in the intra-partum as well as the antepartum period gives valuable information for the management of the high-risk pregnancy.

REFERENCES

1. Miller, DA. Intrapartum fetal heart rate assessment. Lockwood, CJ, ed. UpToDate. Waltham, MA: UpToDate Inc. http://www.uptodate.com (Accessed on August 15, 2017.)
2. Alfirevic Z, Devane D, Gyte GM, Cuthbert A. Continuous cardiotocography (CTG) as a form of electronic fetal monitoring (EFM) for fetal assessment during labour. Cochrane Database Syst Rev 2017; 2:CD006066.
3. Macones GA, Hankins GD, Spong CY, et al. The 2008 National Institute of Child Health and Human Development workshop report on electronic fetal monitoring: update on definitions, interpretation, and research guidelines. Obstet Gynecol 2008; 112:661.
4. Martin CB Jr, de Haan J, van der Wildt B, et al. Mechanisms of late decelerations in the fetal heart rate. A study with autonomic blocking agents in fetal lambs. Eur J Obstet Gynecol Reprod Biol 1979; 9:361.
5. Itskovitz J, LaGamma EF, Rudolph AM. Heart rate and blood pressure responses to umbilical cord compression in fetal lambs with special reference to the mechanism of variable deceleration. Am J Obstet Gynecol 1983;147:451.
6. Electronic fetal heart rate monitoring: research guidelines for interpretation. National Institute of Child Health and Human Development Research Planning Workshop. Am J Obstet Gynecol 1997;177:1385.
7. Johnson TR Jr, Compton AA, Rotmensch J, et al. Significance of the sinusoidal fetal heart rate pattern. Am J Obstet Gynecol 1981;139:446.
8. Jackson M, Holmgren CM, Esplin MS, et al. Frequency of fetal heart rate categories and short-term neonatal outcome. Obstet Gynecol 2011; 118:803.
9. Rouse DJ, Owen J, Goldenberg RL, Cliver SP. Determinants of the optimal time in gestation to initiate antenatal fetal testing: a decision-analytic approach. Am J Obstet Gynecol 1995;173:1357.
10. Practice bulletin no. 145: antepartum fetal surveillance. Obstet Gynecol 2014;124:182.
11. Hoskins IA, Frieden FJ, Young BK. Variable decelerations in reactive nonstress tests with decreased amniotic fluid index predict fetal compromise. Am J Obstet Gynecol 1991;165:1094.

Abnormal CTG: When to Deliver

K Muhunthan, Sir Sabaratnam Arulkumaran

Obstetrics is unique among medical specialties, as it deals simultaneously with two individuals and adverse outcomes for either mother or baby can lead to life-long disability or death. In the event of an adverse outcome, obstetric medicolegal issues become a major burden with high cost to the health system of a country due to financial compensation for families who have to suffer the emotional distress and care for a possibly life-long handicapped child.[1]

However, parents of a handicapped child would any time prefer a normal child than the millions given for child care. Litigation can also be an unpleasant experience and have significant long-term consequences for the working lives for the healthcare staff involved.

Report published by NHSLA in 2015/2016 provides an analysis of the various clinical situations that have led to maternity claims. Confidential inquiries into perinatal deaths and cases of litigation suggests the following key points for litigation:[2]

a. Failure to incorporate the clinical situation
b. Inability to interpret the CTG
c. Delay in taking action
d. Poor communication.

Need to Incorporate Clinical Situation in Interpretation of CTG

The clinical situation of parity, current cervical dilatation, rate of progress, position and station of the presenting part are important in clinical decision-making. Special attention should be paid to monitoring fetuses with the following as they are at increased risk of developing acidosis with a given CTG trace compared with an appropriately grown fetus with clear amniotic fluid.[3]

a. Post-term pregnancy
b. Pre-term pregnancy
c. Intrauterine growth restricted fetus
d. Heavily meconium stained and scanty fluid
e. Intrauterine infection
f. Intrapartum bleeding.

Some medical interventions/conditions may also give rise to fetal compromise:

a. Injudicious use of oxytocin infusion
b. Provision of epidural in late first and second stage with inadequate monitoring of the fetus
c. At the time of delivery of a macrosomic baby with shoulder dystocia, breech delivery, difficult instrumental delivery
d. Abruption, scar rupture and cord prolapse
e. Suspicious or abnormal pathological admission CTG.

Interpretation of the CTG

There are four essential features in interpretation of the fetal heart rate (FHR) trace recorded by electronic fetal monitors:

1. The **baseline FHR** of a term fetus is 110 to 160 bpm and can be read from the CTG

recording. Each fetus will exhibit its own baseline rate.

2. **Accelerations** are sudden rise of the FHR from the baseline for >15 beats for a duration of >15 seconds and are associated with a fetus that is not acidotic.

Accelerations are usually associated with fetal movements and two such accelerations in a 15 minutes period CTG trace is termed a reactive trace. It indicates the presence of a non-hypoxic fetus and the integrity of the 'somatic or voluntary nervous system', i.e. centers that control fetal movements.

3. **Decelerations** are sudden fall of the baseline rate for >15 bpm for >15 seconds. The shapes of the decelerations vary as well as their relationship to contractions. Decelerations indicate a transient stress to the fetus and based on the shape and timing of decelerations to the contractions one could identify the cause of the stress.

   a. **Early decelerations** are "mirror images" of contractions, i.e. the onset and offset of the decelerations mirror that of the onset and offset of the contractions and they are usually <40 bpm in-depth. They are reflective of head compression.

   b. Precipitous fall and quick recovery of the FHR that varies in shape and size with each contraction are due to cord compression and are termed **variable decelerations** as they vary in shape, size and occurrence in relation to the contractions. They have a slight increase in the FHR just before and just after the deceleration due to baro receptor mediated cardiac activity and are described as shouldering.

   c. The decelerations that start after the contractions have commenced are termed **late decelerations** (lag time of >20 seconds between the onset of the contraction to the onset of the deceleration or acme of the contraction to the nadir of the deceleration) and are due to reduced retroplacental perfusion that affects fetal oxygenation.

4. **Baseline variability** is the "wiggliness" of the baseline. The ascending limb is due to the sympathetic and descending limb is due to the parasympathetic activity of the baby's autonomic nervous system. The baseline variability is assessed by measuring the bandwidth of the "wiggliness" seen in one-minute segment of the FHR trace. The normal baseline variability is 5–25 bpm. When it is <5 bpm it is reduced baseline variability. If it is markedly reduced where one could not determine the baseline variability, it is termed 'silent pattern' or a 'flat baseline'. The baseline variability may be reduced due to fetal sleep, drugs that act on the central nervous system, hypoxia, and infection. At times it is due to congenital malformation of the brain or heart, or cerebral hemorrhage. Rarely it is >25 bpm due to overdrive of the sympathetic and parasympathetic activity and is called 'saltatory' pattern and is due to transient hypoxic stress.

Analysis of CTG is based on classifying the four features defined above.

Tracings are classified into one of three classes: normal, suspicious or pathological according to the criteria presented in Table 11.1.[4]

"Cycling" due to fetal behavioural state

Periods of **reactive** segments (two **accelerations** in a 15 minutes segment) with good base line variability (called "active sleep") alternates with non-reactive segments with reduced baseline variability (called "Quiet sleep") and are known as high and low variability cycles or "Cycling". Such pattern is indicative of normal behavioural state of a non-hypoxic fetus at term, which is unlikely to have had a neurological injury in the past.

The fetus with non-reactive CTG for >90 minutes (and also had no previous reactivity) and reduced baseline variability may be hypoxic or infected. If there is no cyclicity it may suggest the possibility of prior

	Normal	Suspicious	Pathological
	Table 11.1: CTG classification criteria, interpretation and recommended management. The presence of accelerations denotes a fetus that does not have hypoxia/acidosis, but their absence during labour is of uncertain significance.		
Baseline	110–160 bpm	Lacking at least one characteristic of normality, but with nopathological features	<110 bpm >160 bpm
Variability	5–25 bpm		Reduced variability for >50 minutes, increased variability for >30 minutes, or sinusoidal pattern for >30 minutes
Decelerations	No repetitive* decelerations		Repetitive* late or prolonged decelerations during >30 minutes or 20 minutes if reduced variability, or one prolonged deceleration with >5 minutes
Interpretation	Fetus with no hypoxia/acidosis	Fetus with a low probability of having hypoxia/acidosis	Fetus with a high probability of having hypoxia/acidosis
Clinical management	No intervention necessary to improve fetal oxygenation state	Action to correct reversible causes if identified, close monitoring or additional methods to evaluate fetal oxygenation.	Immediate action to correct reversible causes, additional methods to evaluate fetal oxygenation (Chapter 4), or if this is not possible expedite delivery. In acute situations (cord prolapse, uterine rupture or placental abruption) immediate delivery should be accomplished.

*Decelerations are repetitive in nature when they are associated with more than 50% of uterine contractions.[5]

neurological injury especially if there is absent baseline variability with or without repeated shallow late decelerations, i.e. ominous CTG trace as shown in Fig. 11.1).

Pathological CTG Patterns

When fetal hypoxia/acidosis is anticipated or suspected (suspicious and pathological tracings), action is required to avoid adverse

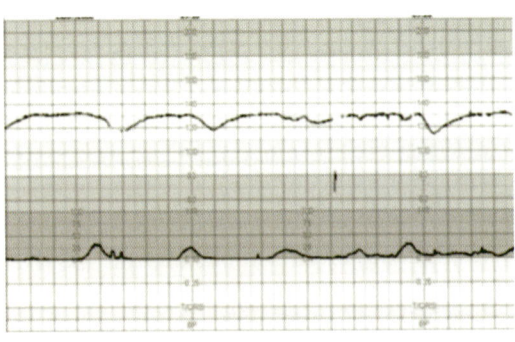

Fig. 11.1: Ominous CTG trace.

neonatal outcome. This does not necessarily mean an immediate cesarean section or instrumental vaginal delivery. The underlying cause for the appearance of the pattern can frequently be identified and the situation reversed, with subsequent recovery of adequate fetal oxygenation and the return to a normal tracing.

The speed of onset and progression of hypoxia can be predicted by the different types of CTG. It is also possible to predict the type of neurological injury based on the type of CTG that gives rise to hypoxia. The discussion below explains the issues.

There are few key CTG patterns that are recognised to be associated with fetal compromise.[6]

In the presence of such pathological CTGs how long can one wait before intervention and does the delay worsen the injury could be answered to some extent by reviewing the patterns discussed follow.

The CTG's could be grouped into those that represent:
1. Acute hypoxia
2. Subacute hypoxia
3. Gradually developing hypoxia and
4. Long standing or chronic hypoxia.

Acute hypoxia presents with profound deceleration with a heart rate < 80 bpm. The pH can drop by 0.01/minute. In addition, there is impact of coronary filling as well. Acute hypoxia is due to reversible (hypotension due to epidural, vaginal examination, artificial rupture of membranes and, at times, no cause may be identified) and nonreversible causes (abruption, cord prolapse, scar rupture). The outcome of the fetus/newborn would depend on the physiological reserve of the fetus, the actual heart rate (whether it is 40 bpm or 60 bpm), the duration of the prolonged deceleration before delivery and the cause for the prolonged deceleration. An example of prolonged deceleration or bradycardia is given below (Fig. 11.2).

If prolonged it can cause fetal death or if born asphyxiated it may lead to acute profound hypoxia and neurological injury around the thalamus and basal ganglia region which may lead to athetoid or dyskinetic type of cerebral palsy.

To optimize the outcome a combination of interventions may be needed namely intrauterine resuscitation, assessment of the stage and progress of labour and prompt delivery of the fetus in the absence of subsequent improvement.

Recommended components of intrauterine resuscitation include:
a. Stopping oxytotics if there is evidence of uterine hyperstimulation. There is a place for use of tocolytics if this does not improve the CTG to improve the uteroplacental circulation.
b. Maternal repositioning to relieve aortocaval compression and cardiac output.
c. High flow oxygen administration to mother if maternal cardiac arrest or abruption of the placenta.
d. Intravenous fluid administration to improve uterine blood flow and to dilute oxytocin in the event of hyperstimulation of the uterus.
e. Tocolysis if there is evidence of uterine hyperstimulation even after stopping oxytotics.

But if the FHR does not return to near normality, operative delivery is advised unless spontaneous vaginal delivery is imminent. Preparation for cesarean or instrumental delivery should be undertaken within six minutes if the CTG was abnormal prior to the bradicardia/prolonged deceleration or if there is evidence of clinical compromise (e.g. IUGR, thick meconium). In the others preparation for delivery should be undertaken within nine minutes if appropriate conservative measures do not reverse the bradycardia/prolonged deceleration in order to prevent long-term neurological injury to the fetus.

Subacute hypoxia presents with prolonged decelerations—The FHR is below the baseline rate for a longer time (e.g. > 60 to 90 seconds) than at the baseline rate (< 30 seconds). With such FHR there is less than optimal circulation through the placenta over a given time especially if the FHR drops to < 80 bpm. With such a trace (Fig. 11.3), some of the fetuses would get compromised with progression of acidosis of approximately 0.01 every two to three minutes.

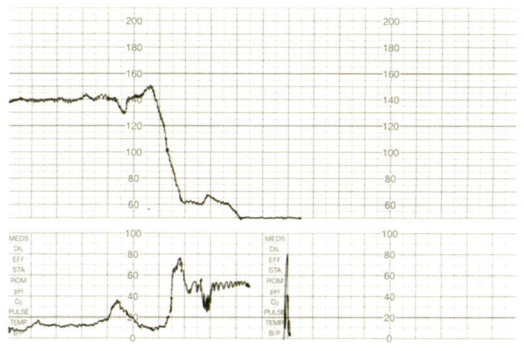

Fig. 11.2: Acute hypoxia.

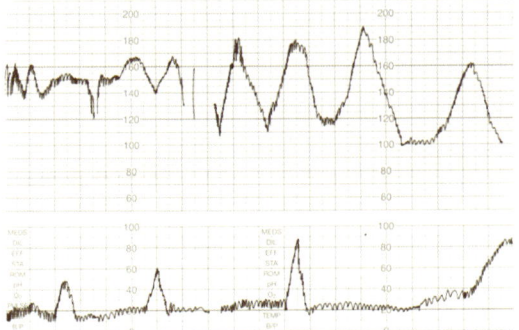

Fig. 11.3: Subacute hypoxia.

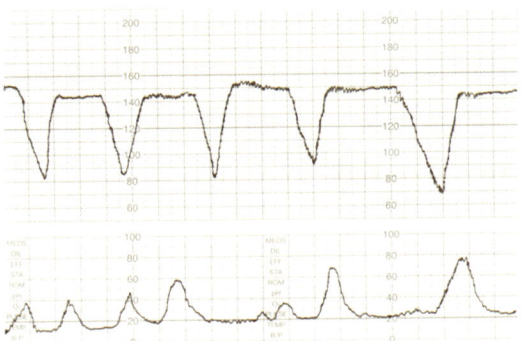

Fig. 11.4: Gradually developing hypoxia.

Subacute hypoxia is usually due to intermittent prolonged cord compression that may be exaggerated by oxytocin infusion or posture. The more the FHR is below baseline FHR and lower the FHR, the higher is the lack of oxygen intake and expulsion of CO_2, i.e. suboptimal circulation which leads to buildup of acidosis.

In subacute hypoxia conservative measures of change of posture, hydration, stopping oxytotics and use of tocolytics should be considered as needed. But if the FHR does not return to near normality, operative delivery is advised unless spontaneous vaginal delivery is imminent. Cesarean or instrumental delivery should be undertaken within 30–40 minutes depending on the stage of labour.

Gradually developing hypoxia: The CTG trace usually starts with a normal baseline rate, normal baseline variability, accelerations and no decelerations. Once decelerations start due to cord compression (variable decelerations) or reduced placental reserve (late decelerations), hypoxia can set in leading to catecholamine surge and rise in the baseline rate. With increasing hypoxia the accelerations do not appear and the decelerations get deeper and wider (i.e. longer duration).

The FHR reaches a peak rate beyond which it is unable to increase the FHR. Even with this rate if oxygenation to the autonomic system cannot be maintained the baseline variability tends to get gradually reduced to almost flat baseline variability. Within one to two hours the fetus may become acidotic. When acidosis gets worse, within a short period the heart rate comes down and becomes asystolic and may end as a stillbirth. If delivered at the 'peak' heart rate within one to two hours of the FHR baseline variability becoming 'flat' ('distress platform') the baby may be born asphyxiated (hypoxia in the tissues and metabolic acidosis) and may suffer neurological injury of bilateral cortical injury leading to cerebral palsy with spastic quadriparesis due to prolonged partial hypoxia. The following CTG (Fig. 11.4), gives an example of gradually developing hypoxia that may lead to as asphyxiated baby.

In gradually developing hypoxia, delivery should be undertaken when there is maximal rise in baseline rate with increasing depth and duration of decelerations with reduction of interdeceleration intervals and marked reduction in baseline variability for a period of one hour unless fetal scalp pH shows that further observation is safe.

Long standing or chronic hypoxia: A fetus that is hypoxic and with early evidence of acidosis may show a nonreactive trace (no accelerations) with absent or markedly reduced baseline variability and shallow decelerations (blunt CNS response) to hypoxic stress produced by contractions (Fig. 11.5).

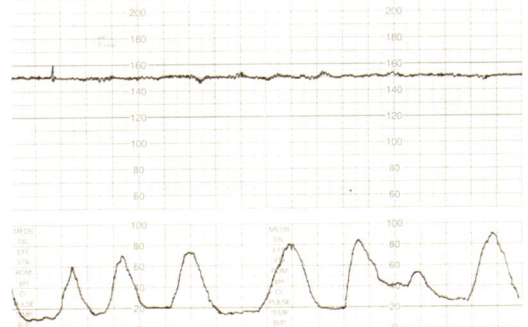

Fig. 11.5: Long standing or cronic hypoxia.

Not all the fetuses may be neurologically affected, and with such trace, earlier the delivery, better the outcome.

A fetus with such a trace may withstand the stress of contractions for variable number of hours without any change on the CTG trace before the fetus gets severely acidotic and has cardiovascular collapse.

Other markers of chronic hypoxia and fetal compromise may be observed in these fetuses, including reduced fetal movement prior to admission, oligohydramnios, presence of old meconium staining of amniotic fluid, meconium aspiration syndrome and subsequent pulmonary hypertension.[7]

In long standing hypoxic patterns, early delivery (40–60 minutes) should be undertaken in the presence of significant meconium, absence of fetal movements, bleeding per vagina, IUGR or prolonged pregnancy. In the absence of such symptoms, observation could be continued for up to 90 minutes before consideration of delivery.

Points to Remember

Key points in interpretation of CTG and optimising management.
1. Accelerations and baseline variability are the hall mark of a nonhypoxic fetus.
2. Better understanding of the significance of the CTG patterns based on pathophysiology is

Contd...

Contd...

important and can be achieved through varied learning modalities.
3. Management decision should be based on the clinical situation in addition to the CTG which may require a senior person to review CTGs when in doubt.
4. Whenever possible an appropriate adjunct method-fetal ECG or fetal scalp blood sampling must be used to confirm the fetal condition without compromising the fetus.
5. Appropriate and timely intervention when conservative measures do not reverse the CTG pattern.
6. Staffing, communication and categorization of the urgency in the event of cesarean section are addition factors in achieving the required decision to delivery time thus improving the fetal outcome.

REFERENCES

1. Williams B, Arulkumaran S. Cardiotocography and medico-legal issues.Best Pract Res Clin Obstet Gynaecol. 2004;18(3):457–66.

2. Resolve and learn. NHS Litigation Authority Annual report and accounts 2015-2016.

3. Muhunthan K, Arulkumaran, S. Medico-legal Issues with CTG. In Chandraharan E (Ed.), Handbook of CTG Interpretation: From Patterns to Physiology. Cambridge University Press 2017;167–70. doi:10.1017/9781316161715.032.

4. Diogo Ayres-de-Campos, Catherine Y Spong, Edwin Chandraharan, FIGO Consensus Guidelines on Intrapartum Fetal Monitoring 2015.

5. Cahill AG, Roehl KA, Odibo AO, Macones GA. Association and prediction of neonatal acidemia. Am J Obstet Gynecol 2012;207:206.e1–8.

6. Arulkumaran S. Fetal surveillance in labour. Munro Kerr's operative obstetrics. Twelfth edition. TF Baskett, AA Calder and Arulkumaran S (Eds) Saunders Elsevier, Edindrough. 2014; 41–56.

7. Ugwumadu A. Recognition of Chronic Hypoxia and the Preterminal Cardiotocograph. In Chandraharan E (Ed.), Handbook of CTG Interpretation: From Patterns to Physiology. Cambridge: Cambridge University Press 2017; 96–100 doi:10.1017/9781316161715.020.

Effects of Various Drugs on CTG

Muralidhar V Pai, Rekha Upadhya

Evaluation of fetal well-being using electronic Fetal Heart Rate (FHR) monitoring needs careful observation and interpretation. A comprehensive description of cardiotoco-grpahy (CTG) consisting of uterine contractions, heart rate, beat to beat variability, presence or absence of accelerations (importance in labour is downgraded now) and decelerations, is needed to contemplate prompt action. Clinician should be aware of the effect of various procedures and drugs used in different medical complications of pregnancy and during labour for augmentation or analgesia. The effects will be discussed with respect to different components of CTG for better understanding.

Baseline Fetal Heart Rate

Normal baseline heart rate is 110 to 160 bpm. Abnormal baseline FHR of more than 160 bpm is called tachycardia and a baseline of less than 110 bpm is called bradycardia.[1,2] Maternal pulse should be measured and documented at the start of every EFM trace.

Fetal Tachycardia

Fetal tachycardia represents increased sympathetic activity or a decrease in parasympathetic activity. Fetal tachycardia can be an early sign of hypoxia, where the fetus is trying to elevate its heat rate to compensate. In preterm fetuses the base line rate will be higher than the term fetus. The common causes of fetal tachycardia are maternal pyrexia, hypoxia and chorioamnionitis and drugs. Drugs causing fetal tachycardia include Beta sympathomimetics (terbutaline, isoxsuprine, ritodrine) and nifedipine.[3-6]

Fetal Bradycardia

Fetal bradycardia is a response of increased vagal tone or a sign of profound myocardial depression which is a preterminal event before fetal death. This can also be seen in cardiac arrhythmias like complete heart block. Bradycardia is almost due to fetal hypoxia when associated with decelerations or a decrease in baseline variability. Other causes of fetal bradycardia are maternal hypothyroidism, beta blockers, maternal hypothermia and prolonged hypoglycemia.

Baseline Variability

The single most important determinant of fetal well-being is baseline variability of a fetal heart rate trace. This cannot be picked up by intermittent auscultation. Decreased baseline variability is seen in preterm fetus or in conditions known to be associated with central nervous system depression. Fetal hypoxia is the most important cause. The commonly used drugs causing central nervous system depression are sedatives like tramadol hydrochloride, pethidine, fantanyl.

The effect of a drug on baseline variability is evident when there is a sudden decrease after the drug administration, especially in the presence of a normal baseline and absence of decelerations. Corticosteroids, magnesium sulphate administration may sometimes cause remarkable decrease in baseline variability. In labour, it is important to differentiate the drug effect to prevent an unnecessary operative intervention. The differentiating parameter is to look for any presence of decelerations for diagnosis of hypoxia as drugs do not cause decelerations. In mothers with epidural labour analgesia, decelerations may be result of hypotension, identification and correction may get the trace normal. Drugs like narcotics (fantanyl) which is used as a part of low dose mixture for pain relief, may be associated with decrease in baseline variability making it difficult to rule out fetal hypoxia.[4,5,7–10] Corticosteroids decrease FHR variability mostly with betamethasone and abolishes diurnal fetal rhythm. $MgSO_4$ decreases short-term variability and insignificant decrease in FHR.

Prolonged Decelerations

It is defined as a deceleration lasting for more than 2 minutes. These become pathological if they cross two contractions, i.e. more than 3 minutes. This can be due to prolonged cord compression, cord prolapse, abruption, scar dehiscence/rupture or severe placental insufficiency, supine hypotension or uterine hyperstimulation. Uterine hyperstimulation with prolonged decelerations is often seen in induced or augmented labours and abruption. Cocaine ingestion was found to have vasospasm resulting in decelerations lasting for more than 3 minutes.

When the duration of deceleration is more than 5 minutes, there may be associated tachycardia and loss of variability due to fetal stress response. Fetal tachycardia with loss of baseline variability and prolonged decelerations may be representing fetal CNS insult and requires prompt resuscitation and removal of the insult.[2–5]

Sinusoidal Pattern

It is a rare, distinct pattern of FHR. This is representative of fetal hypoxia due to fetal anemia and associated with high perinatal morbidity and mortality. These are visually apparent, smooth, sine wave like undulating pattern in electronic FHR baseline with a cycle frequency of 3–5 per minute, which persists for 20 minutes or more. The cause of fetal anemia may be Rh iso-immunization, feto-maternal bleed or fetal hemorrhage. It has also been reported as a manifestation of hypoxia in labour. Drugs like alphaprodine, meperidine and butorphanol have been implicated in the development of sinusoidal pattern in labour. Sinusoidal pattern in the absence of fetal anemia and acidosis may not be significant as it could be physiological due to fetal thumbsucking. Appearance of accelerations in response to stimulation suggests a healthy fetus, thus differentiating from pathological sinusoidal pattern which does not have accelerations.[2,4–6,10]

A derangement of the nervous system due to peripheral or central ischemia is believed to be responsible for sinusoidal pattern. It has been produced in animal models by surgical or chemical vagotomy and the infusion of arginine vasopressin.

Oxytocin and Fetal Heart Changes

Oxytocin is commonly used for induction and augmentation of labour. It does not have direct influence on the FHR or on the controlling cardiac centres in the brain as in the case with some anesthetic and antihypertensive drugs. Its influence is indirect via increased uterine activity, due to increased frequency of contractions or baseline pressure (hypertonus). Increase in duration or amplitude of contractions can also lead to FHR changes. Oxytocin hyperstimulation causes fetal bradycardia due to tetanic or sustained contractions lasting for 3–4 minutes. A rapid decline would be anticipated in post-term and growth restricted fetuses and also with reduced amniotic fluid with thick meconium.

It is advisable to do CTG prior to commencing oxytocin to make sure good fetal health, reflected by normal reactive FHR pattern. If the trace is abnormal then oxytocin should not be used, as it causes further hypoxia to the fetus by reducing the perfusion to the placenta by additional contractions.

When oxytocin is used for augmentation, there may be FHR variations when the dose is increased to achieve the optimal target frequency of contractions. With dose reduction, FHR pattern returns to normal.

Epidural Anesthesia

The insertion of an anesthetic agent into the epidural space can be associated with a degree of instability of the maternal vascular system. Once the circulating volume and vascular stability returns then the trace returns to normal.

Pethidine and Baseline Variability

Pethidine administration may blunt the accelerations and reduce baseline variability as in the sleep phase. In labour hypoxia develops gradually due to regular uterine contractions cutting of blood supply to the placenta. The fetus tends to compensate for the hypoxia by increasing the cardiac output, by increasing the FHR as it has limited capacity to increase the stroke volume. Hence reduction in the baseline variability is likely to be due to Pethidine rather than hypoxia. After the baby is born, the baby may not cry and may need stimulation or assisted ventilation because of the effect of the drug on the central nervous system causing respiratory depression. But the cord arterial blood status indicating that there was intra-uterine hypoxias.

CONCLUSION

One has to be aware of the drugs used and their effects on fetal heart and its variations before interpreting the CTG during antenatal period and particularly in labour to diagnose fetal distress correctly and in time to contemplate appropriate action.

REFERENCES

1. Macones, GeorgeA, et al. The 2008 National Institute of Child Health and Human Development Workshop report on electronic fetal monitoring: Update on definitions, Interpretations, and Research guidelines. Obstet Gynecol 2008;112(3):661–6.
2. American College of Obstetricians and gynecologists. ACOG practice bulletin No. 106: Intrapartum fetal heart rate monitoring: nomenclature, interpretation, and general management principles. Obstet Gynecol 2009;114(1):192–202.
3. Arulkumaran S, Wong YC, Anandkumar C, et al. Sinusoidal like patern associated with acute feto maternal transfusion. Aust NZJ Obstet Gynecol 1989;29:364–5.
4. Freeman RK, Garite TJ, Nageotte MP (Eds). Fetal heart rate monitoring (3rd edn.) Philadelphia: Lippincott Williams and Wilkins, 2003.
5. Gibb D, Arulkumaran S, (Eds.). Fetal monitoring in Practice, 3rd edn. Edinburgh: Churchill Livingstone, 2008.
6. National Institute of Health and Clinical Guidance (NICE). Intarapartum care: Care of healthy women and their babies during childbirth. September 2007. London: National Institute for Health and Clinical Excellance. www.nice.org, uk/CG055.
7. Larma JD, Silva AM, Holcroft CJ, et al. Intrapartum electronic fetal heart rate monitoring and the identification of metabolic acidosis and hypoxic ischemic encephalopathy. Am J Obstet Gynecol 2007;197(3):301.e 1–8.
8. John RM Rodney, Benjamin J FH, MacMillan R. Electronic fetal monitoring. Family medicine Obstetrics. Prim Care Clin Office Pract 2012;39: 115–33.
9. Spong CY. Electronic fetal heart rate monitoring: another look (Editorial). Obstet Gynecol 2008;112: 506–7.
10. Williams B, Arulkumaran S. Cardiotocography and medicolegal issues. Best Pract Res 2004;18: 457–66.

Cardiotocography in Twin Pregnancy

Nuzhat Aziz, Pallavi Chandra R

Twin pregnancy is regarded as one of the high-risk pregnancies, frequently encountered in obstetrics. The proportion of these pregnancies have been increasing due to assisted reproductive techniques.[1] Multifetal pregnancies have higher incidence of maternal and fetal complications, warranting a need for better fetal surveillance. The incidence of preterm labour, congenital anomalies, fetal growth restriction and twin to twin transfusion is increased in twin pregnancy. Twins were found to have four times risk of cerebral palsy.[2] The second fetus is at more disadvantage for fetal monitoring due to technical difficulties and our inability to do fetal blood sampling. Fetal surveillance in twins is required to optimize the timing of delivery to minimise the neonatal morbidity/mortality and also to prevent intrauterine compromise or death. This chapter will focus on the modalities available for cardiotocography (CTG) in twin pregnancies, difficulties and possible solutions.

Chorionicity

The perinatal outcome is largely driven by the chorionicity of the multifetal pregnancy. Fetal growth restriction, twin specific complications and discordant growth all make a fetus more vulnerable to hypoxic changes. Cardiotocography is one of the basic tools for monitoring of twin pregnancy. Monoamniotic twins have the maximum difficulty in surveillance, both

with ultrasound and with Electronic Fetal Monitoring (EFM). These pregnancies are frequently associated with polyhydramnios, increased syncronicity of FHR patterns in comparison to dichorionic twins. The chorionicity of a twin pregnancy should be identified and labelled early in pregnancy for management plans and surveillance.

Differences between a Singleton and Twin Pregnancy Monitoring

The requirements of twin CTG is quite similar to the CTG performed for a singleton pregnancy, provided the machine has dual channel monitoring facility with two ultrasound tranducers. The challenge is to capture both traces simultaneously, minimise signal loss, visualise them separately and then interpret the characteristics. Interpretation to appropriate and timely action is the same as for singleton pregnancy. Misinterpretation and human error in recognising abnormal patterns have been reported as the main reasons for poor outcomes despite the use of continuous electronic fetal monitoring. Twin traces double the probability of these errors, which can only be avoided through continued training and regular credentialling.

CTG TRACE ACQUISITION

The most important aspect of cardiotocography is to obtain a good quality FHR trace.

Simultaneous monitoring of twins with a single machine is preferable to nonsimultaneous monitoring with two machines. Twin CTG machines are commercially available, with two transducers for accurately monitoring and differentiating both twins. The twin trace has special markings and denotations as shown in Fig. 13.1.

The recommended maternal position for cardiotocography is lateral recumbent or semi sitting or upright positions to avoid aortocaval compression. The recommendations for paper speed remain the same as for singletons and as per the institutional protocol. The second step of trace acquisition is localisation of the fetal hearts of twin A and B. It is a good clinical practice to use a bedside ultrasound machine to identify twin A and B and respective fetal heart locations before application of US transducers. The first US transducer must be applied to twin A so that, it is recorded as FHR1, FHR2, FHR3 (in triplets) is denoted by the order of application of the US transducers. If the fetal scalp electrode is applied for twin A after applying external transducers, it may erroneously be traced as DFHR2 as shown in Fig. 13.2.

Antenatal Cardiotocography

While all of us do consider twin pregnancies as high-risk and do start antenatal surveillance with CTG, there are no evidence-based recommendations and proven benefits for antenatal surveillance with cardiotocography in twins. The guidelines on management of multifetal gestations from Royal College of Obstetricians and Gynaecologists or NICE do not have any suggested algorithm for antenatal monitoring. Chorionicity based surveillance was mentioned in the 2014 American College of Obstetrics and Gynecology practice bulletin on antepartum fetal surveillance, with recommendation to start surveillance with weekly NST from 34 weeks for uncomplicated dichorionic twins and from 32 weeks for uncomplicated monochorionic twins. Monoamniotic twins were given the suggestion to start surveillance from 28 weeks. Complicated twins may require surveillance protocols as per the added risk recommendations. The ideal frequency of monitoring again has not been an evidence-based suggestion but commonly recommended at weekly intervals.[3,4] The interpretation of antepartum traces remain the

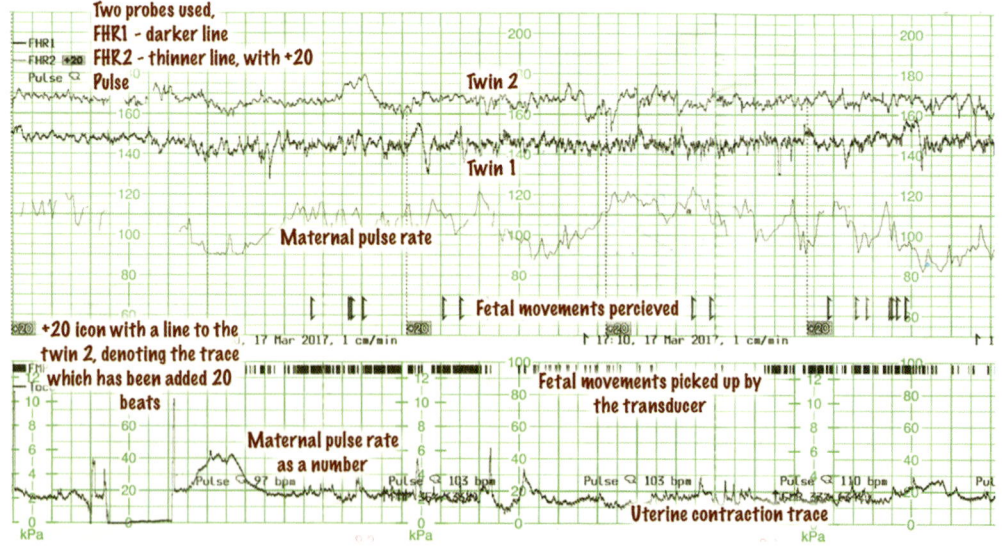

Fig. 13.1: Markings on a twin CTG trace.

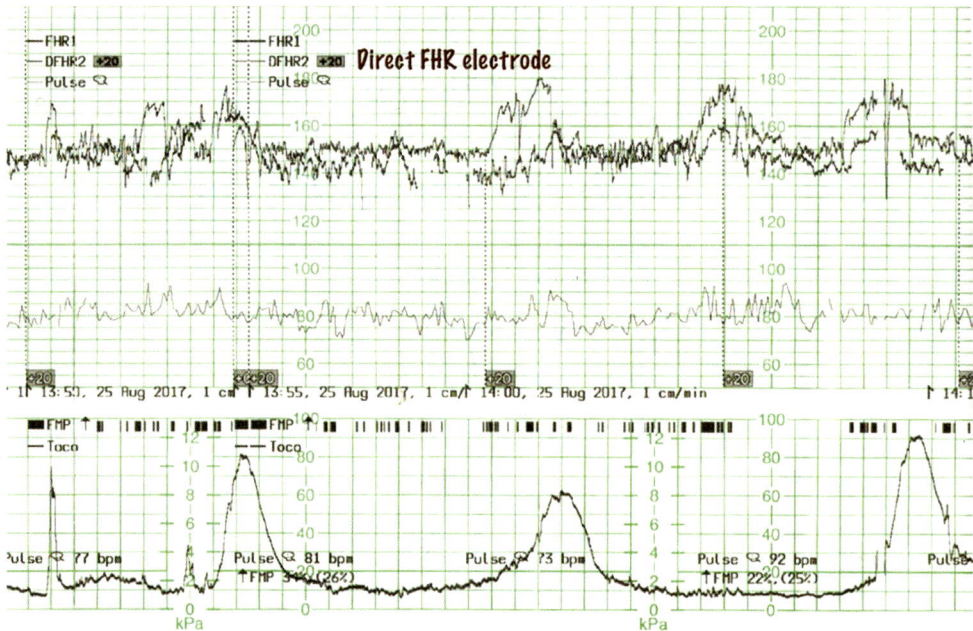

Fig. 13.2: Twin monitoring with fetal electrode for the first twin.

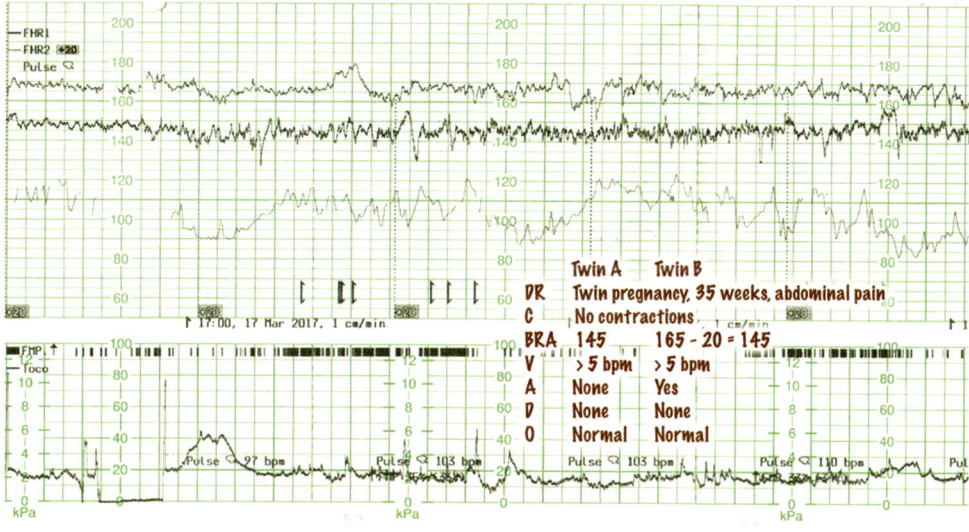

Fig. 13.3: Interpretation of traces using the pneumonic DR C BRAVADO.

same as the singleton pregnancy. Twins were found to exhibit synchronous behaviour patterns in 94% of the time (sleep and awake cycles) as shown in Fig. 13.2. The accelerations may thus be simultaneous, making it difficult to differentiate between the two traces.[5–7]

SYNCHRONY IN CARDIOTOCOGRAPHY

The Nonstress Test (NST) is dependent on accelerations of the FHR commonly associated with fetal movements. It is the most common test used for fetal surveillance in the antepartum period. Studies have reported that twin NSTs show synchrony or similarity in the tracings in about 60% after 28 weeks of gestation. Synchrony may be seen in higher percentage in monochorionic twins, similar sized twins or with fused placentae. Fetal heart rate accelerations were found to be simultaneous in 36%, with synchronous behaviour patterns in 94.7% of the time in a study done in the year 2000.[5-9] This is of great importance for antenatal electronic fetal monitoring with NST for synchronicity with similar accelerations, similar awake and sleep cycles increases the difficulties of inter-pretation of a twin trace.

EXTERNAL OR INTERNAL CARDIOTOCOGRAPHY

Antenatal cardiotocography for twins require the external transducers, similar to a singleton pregnancy. The twin CTG machine has the ability to compare the two recorded traces, differentiate between the two and alert when synchronous. The most common reasons for signal losses are maternal obesity, fetal move-ments, polyhydramnios, receiving signals from the second twin, picking up the maternal pulsations, maternal pushing efforts and sometimes delivery of the first twin.

However when twins are in labour, internal cardiotocography with Fetal Scalp Electrode (FSE) is considered optimal (after amniotic membrane rupture). The fetal electrode has an ECG based capture of the FHR trace, in comparison to the ultrasound based capture of external cardiotocography. Fetal scalp electrode (FSE) has a much lower rate of signal loss even in the second stage of labour. The signal loss is believed to be 5 to 7% in first stage of labour and 15 to 19% in second stage in singleton pregnancies. The signal loss rate for twins external cardiotocography in first stage of labour was estimated to be 26 to 33% and 41 to 63% in second stage.[8] The rates for signal loss in the first and second stage for FSE were 0.9% and 9% respectively.[9]

Wrong application on the cervix, dislodge-ment of the electrode may still be the reasons for signal loss in internal cardiotocography.

As discussed earlier, an important practical point is that the FSE (DFHR - direct FHR) cable should be the first transducer applied to the machine to get a denotation of DHFR1, followed by the external US transducer to the twin B which gets denoted as FHR2. The contraindications to FSE placement remains the same as for any singleton pregnancy— non-vertex presentation, any concern of vertical transmission (Hepatitis A, B, C, E, HIV, active genital herpes), any fetal bleeding disorder or prematurity less than 32 weeks. Maternal heart patterns have also been documented to be picked up in about 5% of the traces.

Differentiation of a Twin Trace

Most machines have an inbuilt identification system to demarcate twin A and B with different shades of grey. The darkest line denotes the FHR1, the first FHR transducer which was used. Sometimes it becomes difficult to identify traces based on the shading of the line only, if the FHR are similar or differ by less than 20 beats per minute (bpm). A very important concept to note is a special feature of the twin CTG machine which allows the addition of 20 beats to the actual FHR of second applied US transducer (Fig. 13.4). This plus 20 option results in the top-most trace belonging to the twin with second transducer (FHR2) as shown in Fig. 13.4. A trace showing FHR2 as 160 bpm, is in actual 140 bpm. If a three channel CTG machine is used for triplets then FHR2 gets +20 added, and FHR3 has −20 subtracted to separate the traces. Care should be taken to understand this concept and prevent misinterpretation of the traces. Practically we apply the transducers, start a trace and use this option if there is overlap of the traces. Assignment of the transducers to each twin is essential to follow FHR changes over a period of time. We should aim to get FHR1 for twin A and FHR2 for twin B. Fetal electrode application should then be DFHR1 (first applied) followed by the second twin

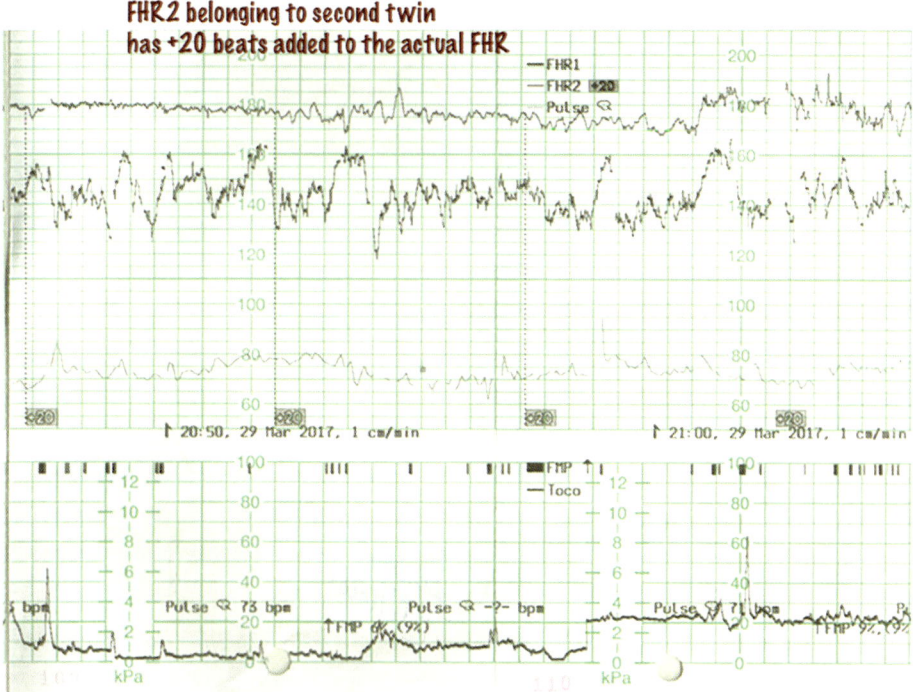

Fig. 13.4: Separation of the twin FHR traces.

FHR2. Failure to follow this sequence will result in jumping of twin traces from A to B and vice versa.

INTRAPARTUM FETAL MONITORING

The use of cardiotocography as a means of detecting hypoxia in labour has become a routine care for all high-risk pregnancies. Even though it has low sensitivity and specificity for detecting hypoxia, all recommendations are for continuous CTG monitoring in the intrapartum stage. The challenge during second-stage labour is to monitor both the twins, without loss of contact, loss of signal, with maximum accuracy and to capture a good quality trace. The FHR parameters and the classifications remain the same as for singleton pregnancies. The FHR changes and their relation to evolving hypoxia, acute onset or long standing hypoxia have to be identified and action taken. Labelling has to be done for two fetuses separately (Fig. 13.4). The first twin allows the possibility of a fetal blood sampling (FBS) to confirm any non-reassuring changes in the CTG whereas FBS cannot be done for the second twin until the first twin is delivered. The indications, classifications and interpretation of FBS remain the same as that of any singleton pregnancy.

Historically, a higher morbidity and mortality for second twins has encouraged physicians to expedite the delivery of the second twin within a specific time frame (30 minutes) after the delivery of the first twin. The significant added risks to the second twin are uterine inertia, malpresentation, umbilical cord prolapse, abruption, and stillbirth. However, obstetric technological advances with continuous FHR monitoring and portable real-time ultrasound are beginning to equalize survival rates for first and second twins. Regardless of the mode of delivery, the first twin has better outcomes with higher pO_2 and lower pCO_2. This finding was consistent, regardless of presentation (both vertex,

vertex-nonvertex, or both nonvertex) or mode of delivery (vaginal or cesarean section) and irrespective of the interval between deliveries. Carbon dioxide retention was highest in the group requiring cesarean section for the second twin only.

After the delivery of the first twin, portable real-time ultrasound equipment is of great help in the delivery room. It will allow the physician to rapidly verify the presentation, the FHR, and the status of the umbilical cord of the second twin. Probability of success for intrauterine manipulation or external version can be improved with the support of ultrasound.

Once the second twin settles into the pelvis, an amniotomy and spiral electrode application can be accomplished. Extra equipment, including a hand-held Doppler, should be available so that the physician can hear the FHR until a cardiotransducer or spiral electrode can be applied. Evaluation of uterine activity is critical. Uterine inertia may need to be treated with oxytocin. Correcting the maternal position with a lateral tilt may improve uterine perfusion and function. The transducers should be re-applied after the delivery of the first twin, since the uterine contour and the position of the fetus changes.

Delivery of the second twin by cesarean section is always a possibility. Failed version or extraction of a fetus in transverse or oblique lie, a prolapsed umbilical cord or fetal distress may necessitate a cesarean section. Electronic fetal monitoring should be continued during preparation for the procedure. All necessary equipment and personnel should be anticipated and available during the second-stage of a twin delivery.

SPECIAL SITUATIONS IN TWINS

Twin to twin transfusion can lead to fetal anemia and sinusoidal CTG pattern for one twin (Fig. 13.5). Severe polyhydramnios–oligohydramnios sometimes pushes one twin to the posterior aspect of the uterus leading to difficulty in picking up the heart rate, the

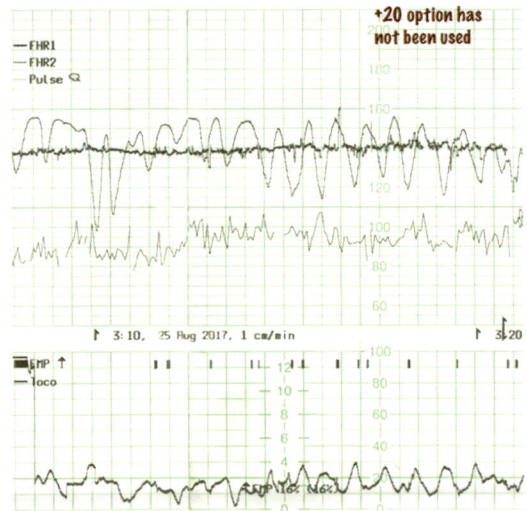

Fig. 13.5: Twin to twin transfusion CTG.

other twin with polyhydramnios offers difficulty due to excessive movements. There may be situations when fetal monitoring with CTG is not possible and we may then be dependent on real–time ultrasound with Doppler studies and biophysical profile to monitor the fetus. Monoamniotic twins are rare but associated with disproportionately high perinatal mortality. Cord entanglement is a unique risk that accounts for half of the deaths in monoamniotic twins. Pseudomonoamniotic twins are those who have ruptured the inter twin membrane either spontaneously or iatrogenically to become monoamniotic. There is no consensus on the ideal time for start of surveillance or type of surveillance or timing of delivery. Cord entanglement related deaths are acute events leading to sudden death; many have considered the theoretical possibility of continuous electronic monitoring of these twins. Studies have shown that successful continuous antenatal CTG monitoring can be done for 50% of the time, and improves with increasing gestational age. Practically, continuous inpatient CTG monitoring is difficult to achieve, affected by many factors, and has the potential for increasing medicolegal liability by establishing a standard of care that cannot be met.

Computerised Decision-Making Support

Human factors have been a reason for misinterpretation, inadequate response to an abnormal CTG. Decision support systems are computerised systems for CTG interpretation that create alerts for a review by the clinican. They have been available since the late 1990s but Cochrane review in 2010 suggested a better outcome with the use of computerised CTG in antenatal period with a reduced relative risk for perinatal mortality (RR 0.2, 95% CI 0.04–0.88).[10] Studies using computerised

CTG in the intrapartum period also suggested that it had a scope for better performance; was found to be as good as experts in interpretation, and had an earlier decision-making. The infant trial was a largest multicentric randomised control trial designed to study this effect; with 46,000 women enrolled after 35 weeks gestation; both singleton and twin pregnancies. The results have not shown any differences in outcomes between the two groups with or without decision-making support.[11]

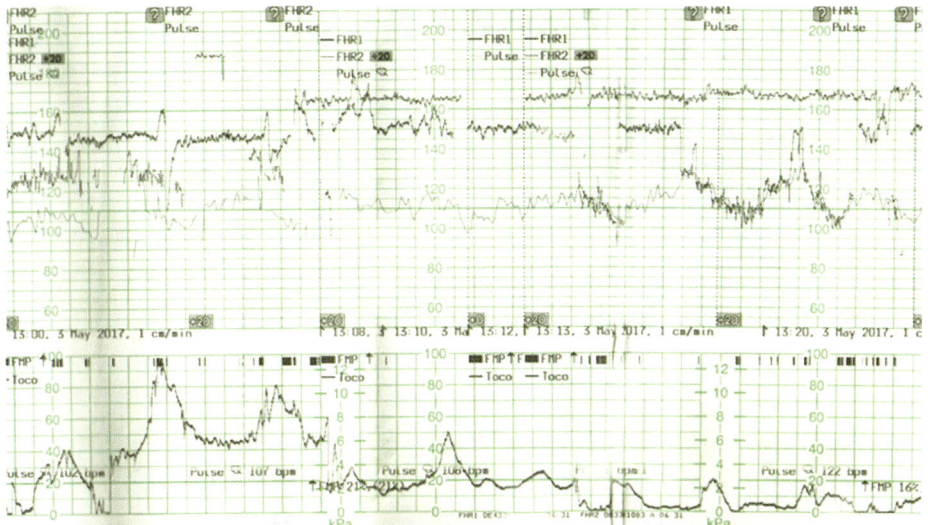

Fig. 13.6: Case CTG 1.

	CTG 1300 to 1325 hours	Twin 1	Twin 2
	Quality of trace is poor, signal loss		
DR	Define risk: DCDA twins, in labour at 3 cm		
C	Contractions: not picking up clearly, in labour	2/10 minute	2/10 min
BRA	Baseline rate	140	130
V	Variability	<5	>5
A	Accelerations	Present	None
D	Decelerations	None	None
O	Opinion	Normal	**Normal**
	A poor quality trace, with loss of signal for both transducers, transfer of signal to second twin or maternal heart rate. The query marks at the top of the trace hints at possible same signal pick-up from two transducers, and prints the names of the transducers. Assessment of the quality of trace should be a prerequisite before using DR C BRAVADO for interpretation (Fig. 13.3).		

Example of Intrapartum Monitoring of Twins with its Difficulties (Figs 13.6 to 13.12)

Primigravida, 28 years of age, BMI 32, IVF conception, DCDA twin pregnancy in labour at 36 weeks of gestation. Both fetuses were cephalic, appropriate for gestational age, with estimated fetal weight of 2.4 and 2.5 kg with placenta anterior upper segment. Vaginal examination showed cervical dilatation of 2 cm with rupture of membranes, clear amniotic fluid with the presenting part vertex at 1 station.

The mnemonic DR C BRAVADO is a very useful tool for a systematic interpretation of a CTG trace. The interpretation of the twin trace will require two separate columns for twin 1 and twin 2, and opinion has to be given for each twin. Management plans applied for one fetus in twin pregnancy has the ability to affect the second twin.

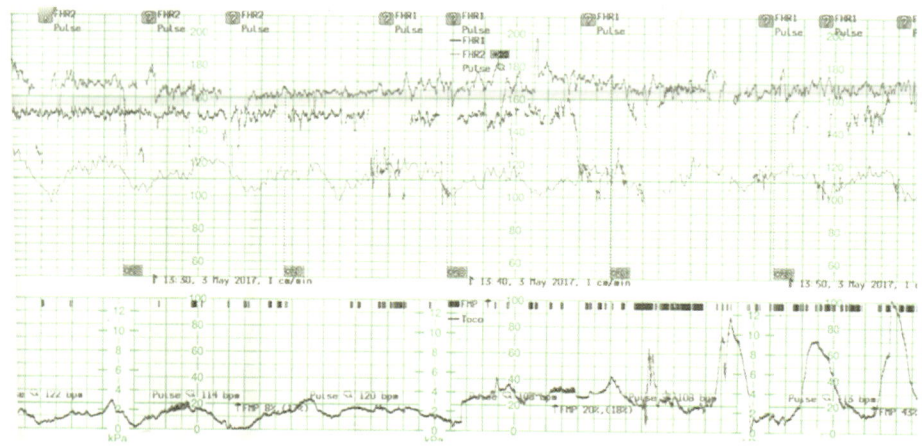

Fig. 13.7: Case CTG 2.

	CTG 1325 to 1355 hours, Cx 3 cm, no progress	Twin 1	Twin 2
	Quality of trace	Poor	Good
DR	Define risk: DCDA twins, in labour		
C	Contractions: not picking up clearly, in labour	3/10 minutes	3/10 minutes
BRA	Baseline rate:	150	145
V	Variability	>5	>5
A	Accelerations	Present	Present
D	Decelerations	None	None
O	Opinion	Normal	Normal
	The first transducer is having signal loss, making it difficult to interpret the CTG. The opinion for the first trace may not be absolutely correct for its difficult to interpret with signal loss. The FHR1 seems to be picking up maternal pulsations, resulting in queries being printed frequently.		

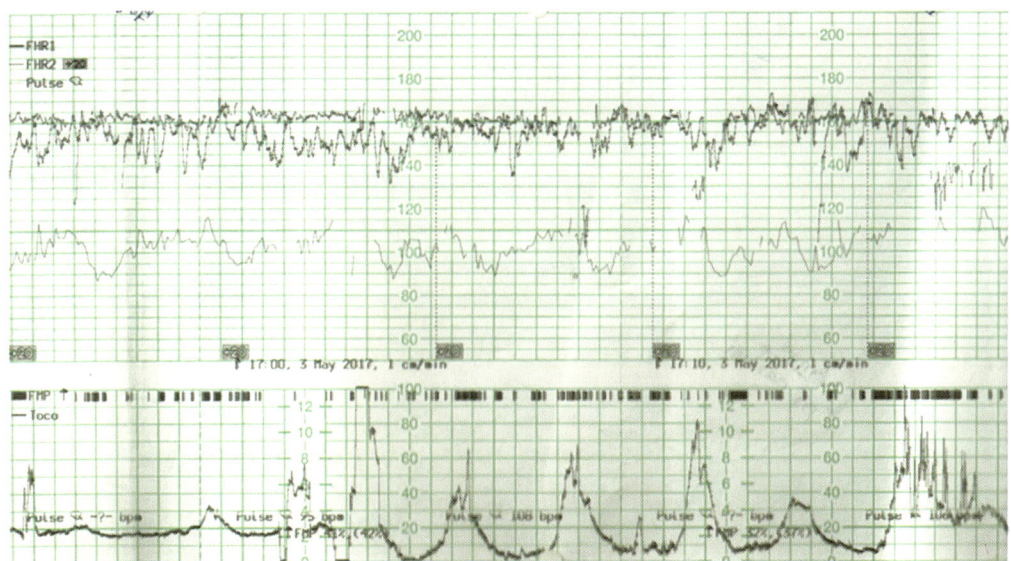

Fig. 13.8: Case CTG 3.

	CTG 1650 to 1715 hours, Cx 6 cm, PPvx 0	Twin 1	Twin 2
	Quality of trace	Good	Good
DR	Define risk: DCDA twins, in active labour		
C	Contractions	4/10 minutes	
BRA	Baseline rate	150	145
V	Variability	> 5	> 5
A	Accelerations	None	None
D	Decelerations	None	None
O	Opinion	Normal	Normal
	The signal pick-up is good for both the traces but we are having difficulty in separating the traces from each other, despite the +20 tool use. One option is to remove the +20 option and check or to put up the fetal electrode for the presenting twin.		

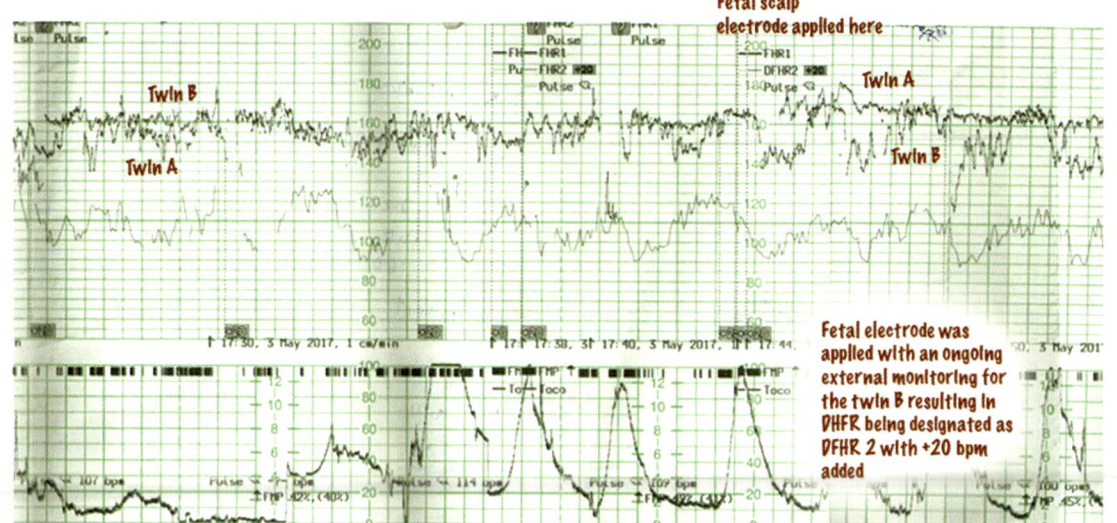

Fig. 13.9: Case CTG 4.

	CTG 1725 to 1755 hours, Cx 6 cm, PPvx 0	Twin 1	Twin 2
	Quality of trace	Good	Good
DR	Define risk: DCDA twins, in active labour		
C	Contractions	4/10 minutes	
BRA	Baseline rate	150	145
V	Variability	> 5	> 5
A	Accelerations	None	None
D	Decelerations	None	None
O	Opinion	Normal	Normal
	The signal pick-up is good for both the traces but we are having difficulty in separating the traces from each other, despite the +20 tool use. One option is to remove the +20 option and check or to put up the fetal electrode for the presenting twin.		

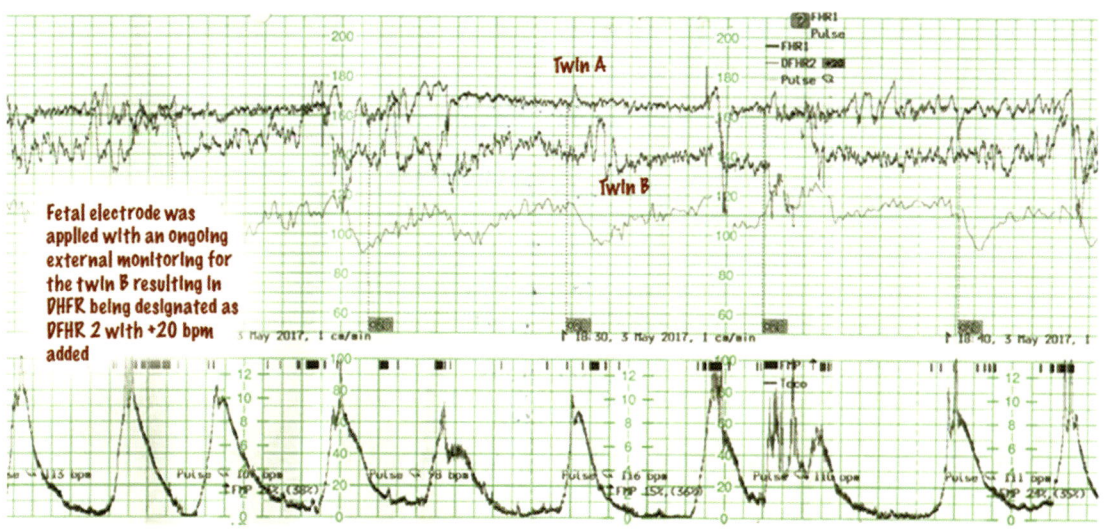

Fetal electrode was
applied with an ongoing
external monitoring for
the twin B resulting in
DHFR being designated as
DFHR 2 with +20 bpm
added

Fig. 13.10: Case CTG 5.

	CTG 1815 to 1845 hours, fully dilated at 1840, PPvx 0	Twin 1	Twin 2
	Quality of trace	Good	Good
DR	Define risk: DCDA twins, in active labour		
C	Contractions	3/10 minutes	
BRA	Baseline rate	145	140
V	Variability	> 5	> 5
A	Accelerations	None	None
D	Decelerations	None	None
O	Opinion	Normal	Normal
	The signal pick-up is good with adequate separation of traces even in the second stage, allowing for appropriate interpretation of trace.		

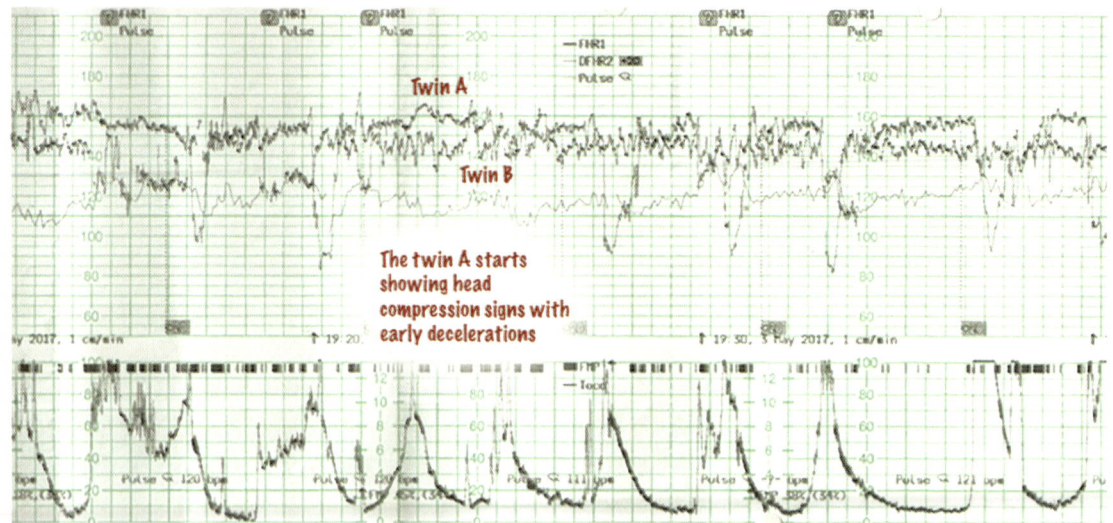

Fig. 13.11: Case CTG 6.

	CTG 1910 to 1940 hours, fully dilated at 1840, PPvx 0	Twin 1	Twin 2
	Quality of trace	Good	Good
DR	Define risk: DCDA twins, in active labour		
C	Contractions	4/10 minutes	
BRA	Baseline rate	135	140
V	Variability	> 5	> 5
A	Accelerations	None	None
D	Decelerations	Early decelerations	None
O	Opinion	Normal	Normal
	The decelerations can be made out very clearly, care has to be taken to interpret which twin is having the deceleration. The screen display of the machine will make it easier to identify the twin showing decelerations.		

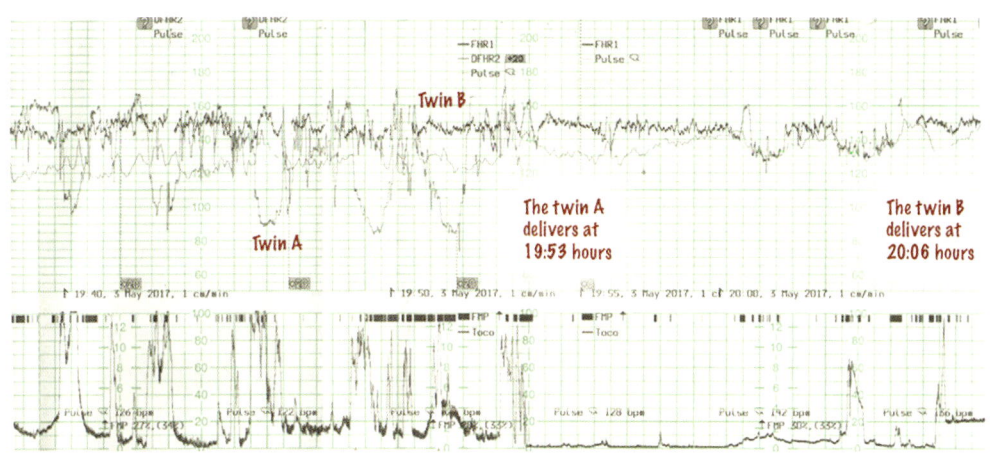

The twin A delivers at 19:53 hours

The twin B delivers at 20:06 hours

Fig. 13.12: Case CTG 7.

	CTG 19:40 to 20:06 hours, fully dilated at 1840	Twin 1	Twin 2
	Quality of trace	Good	Good
DR	Define risk: DCDA twins, in active labour		
C	Contractions	4 /10 minutes	
BRA	Baseline rate	135	150
V	Variability	> 5	> 5
A	Accelerations	None	None
D	Decelerations	Variable decelerations >60 seconds >60 beats falls	None
O	Opinion	Non reassuring	Normal
	The decelerations which seem to touch 90 bpm are +20, so the baseline is 135 bpm with decelerations up to 70 bpm. With the delivery of the twin A, the fetal electrode is removed and the second twin trace continues as FHR1 through external transducer. It should be noted that maternal tachycardia is almost 140/min and creates alerts for it is similar to FHR.		

CONCLUSIONS

Cardiotocography has become a standard fetal surveillace tool in antepartum and intrapartum care of twins and triplets pregnancies.[12–14] The difficulties of external monitoring are doubled in twins and hence require a focussed attention to the clinical background and interpretation of each trace. The dual channel twin monitors have special features to differentiate the traces, which should be used.[15] The availability, knowledge and experience with internal FHR monitoring (FSE) is essential for optimal fetal monitoring of a twin pregnancy in labour. Vaginal twin deliveries can have good outcomes with training in cardiotocography and intrapartum procedures.[16]

REFERENCES

1. Fell DB, Joseph K. Temporal trends in the frequency of twins and higher-order multiple births in Canada and the United States. BMC Pregnancy Childbirth 2012;12:103.
2. Hack KE, Derks JB, Elias SG, Franx A, Roos EJ, Voerman SK. Increased perinatal mortality and morbidity in monochorionic versus dichorionic

twin pregnancies: clinical implications of a large Dutch cohort study. BJOG 2008;115(1):58-67.

3. Hogle KL, Hutton EK, McBrien KA, Barrett JF, Hannah ME. Cesarean delivery for twins: a systematic review and meta-analysis. Am J Obstet Gynecol 2003;188(1):220–7.

4. American College of Obstetricians and Gynecologists. Practice Bulletin No. 145. Antepartum fetal surveillance. Obstet Gynecol 2014;124: 182–92.

5. Pernoll ML, Carnes RW. Electronic fetal monitoring of twin gestations. Am J Obstet Gynecol 1973;116(4):583–4.

6. Bernardes J et al. Fetal heart rate baselines in twins. Interobserver agreement in antepartum estimation. J Reprod Med 2000;45(2):105–8.

7. Gallagher MW, Costigan K, Johnson TR. Fetal heart rate accelerations, fetal movement, and fetal behavior patterns in twin gestations. Am J Obstet Gynecol 1992;167(4 Pt 1):1140–4.

8. Bakker PC, Colenbrander GJ, Verstraeten AA, Van Geijn HP. Quality of intrapartum cardiotocography in twin deliveries. Am J Obstet Gynecol 2004;191(6):2114–9.

9. Ohel G, Samueloff A, Navot D, Sadovsky E. Fetal heart rate accelerations and fetal movements in twin pregnancies. Am J Obstet Gynecol 1985; 152(6 Pt 1):686–7.

10. Grivell RM, Alfirevic Z, Gyte GML, Devane D. Antenatal cardiotocography for fetal assessment. Cochrane Database of Systematic Reviews 2015; Issue 9. Art. No.: CD007863. doi: 10.1002/14651858. CD007863.pub4.

11. Brocklehurst P, on behalf of The Infant Collaborative Group. A study of an intelligent system to support decision making in the management of labour using the cardiotocograph–the INFANT study protocol. BMC Pregnancy and Childbirth 2016;16:10. doi:10.1186/s12884-015-0780-0.

12. American College of Obstetricians and Gynecologists, Society for Maternal-Fetal Medicine. ACOG Practice Bulletin No. 144: Multifetal gestations: twin, triplet, and higher-order multifetal pregnancies. Obstet Gynecol. 2014;123 (5):1118–32.

13. National Institute of Health and Clinical Excellence; Clinical Guideline Number 129. Management of Twin and Triplet Pregnancies in the Antenatal Period 2011.

14. American College of Obstetricians and Gynecologists. Practice Bulletin No. 144. Multifetal gestations: twin, triplet, and higher-order multifetal pregnancies. Obstet Gynecol 2014;123: 1118–32.

15. Avalon fetal monitor FM20/30, FM40/50 by Philips Training Guide, Release G.0 with software revision G.02.xx; 2010, printed in Germany.

16. Rzyska E, Ajay B, Chandraharan E. Safety of vaginal delivery among dichorionic diamniotic twins over 10 years in a UK teaching hospital. Int J Gynaecol Obstet 2017;136(1):98–101. doi: 10.1002/ijgo.12017. Epub 2016 Nov 3.

Interpretation of Abnormal Trace and its Management

G Selvanandhini, Mirudhubashini Govindarajan

The primary aim of cardiotocographic monitoring is to assess the adequacy of fetal oxygenation and presence of fetal metabolic acidemia during labour. This helps in timely intervention and to reduce the risk of neurologic injury or death.

Interpretation of CTG includes a qualitative and a quantitative assessment of the baseline FHR, variability, presence/absence of accelerations, decelerations, change in FHR pattern over time. Each one of these factors individually or collectively can influence the management decisions. The International Federation of Gynecology and Obstetrics (FIGO) consensus guidelines on intrapartum fetal monitoring in 2015[1] has categorised the collective CTG findings as normal, suspicious or pathological.

INTERPRETATION AND MANAGEMENT APPROACH

Bradycardia

Fetal bradycardia is defined as a baseline FHR value below 110 bpm lasting more than 10 minutes (Fig. 14.1).[1]

Causes:

1. Maternal hypotension
2. Tachysystole
3. Uterine rupture or scar dehiscence
4. Placental abruption
5. Cord prolapse
6. Rapid descent of the fetal head.

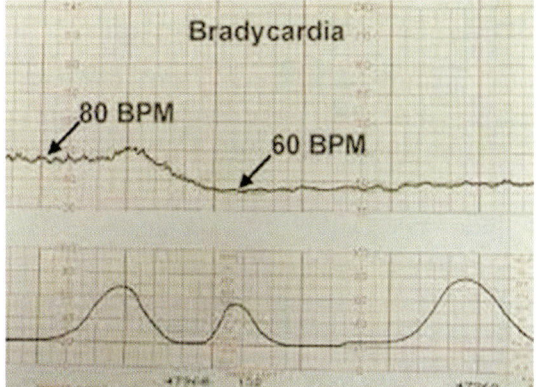

Fig. 14.1: Fetal bradycardia: CTG trace showing a baseline fetal heart rate of 80 bpm later dropping to 60 bpm.

Management

Based on underlying pathology. Evaluation of maternal blood pressure, strength and frequency of uterine contraction, physical examination to rule out placental abruption, uterine rupture, cord prolapse, rapid descent of fetal head should be done. Delivery is indicated if correction of the underlying cause is not possible or if resuscitative measures fail to correct the bradycardia.

TACHYCARDIA

Fetal tachycardia is defined as a baseline FHR greater than 160 bpm for at least 10 minutes (Fig. 14.2).[1]

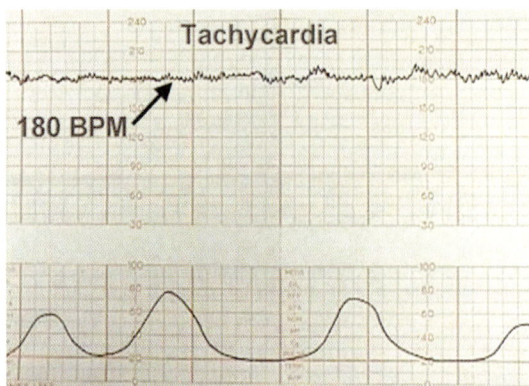

Fig. 14.2: Fetal tachycardia: CTG trace showing a baseline fetal heart rate of 180 bpm.

Causes:

1. Maternal pyrexia
2. Medications (e.g. betamimetics, atropine)
3. Maternal hyperthyroidism
4. Placental abruption
5. Fetal hypoxia
6. Chorioamnionitis
7. Fetal tachyarrhythmia, (such as atrial flutter or supraventricular tachycardia). This rare tachyarrhythmia is characterized by a very high FHR, often in excess of 200 bpm.

Management

Fetal tachycardia alone has not been strongly associated with fetal acidemia, unless associated with recurrent decelerations, absent accelerations, or minimal/absent variability. The evaluation of fetal tachycardia should include assessment for maternal infection or placental abruption and the history of maternal medications. Appropriate treatment should be initiated if the underlying cause can be identified and treated (e.g. paracetamol for reduction of maternal pyrexia and antibiotics for treatment of chorioamnionitis). In addition, fetal scalp stimulation should be performed to provoke FHR acceleration, which is a sign that the fetus is not acidotic. Delivery is indicated if acidemia or placental abruption is suspected.

DECELERATIONS: CAUSES AND MANAGEMENT

Variable Decelerations

Variable decelerations occur due to umbilical cord compression (Fig. 14.3). Occasional variable decelerations are common intrapartum and do not result in adverse consequences when associated with a good variability and accelerations.[1] Transient cord compression is well-tolerated by the fetus and usually does not require any intervention. Recurrent variable decelerations require a close monitoring of the CTG for loss of variability and accelerations.

Management of variable decelerations involves maternal repositioning to relieve the cord compression. Amnioinfusion can be tried for potential or suspected umbilical cord compression and it reduces the occurrence of variable FHR decelerations, reduces the cesarean section rates and improves short-term neonatal outcomes.

Early Decelerations

Early decelerations are drop in FHR with uterine contractions probably because of head compression causing vagal stimulation with heart rate deceleration (Fig. 14.4). Early decelerations are usually benign and are not associated with fetal hypoxia or acidemia

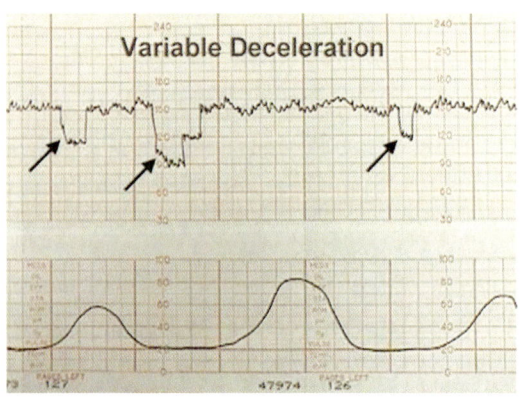

Fig. 14.3: CTG trace showing variable decelerations.

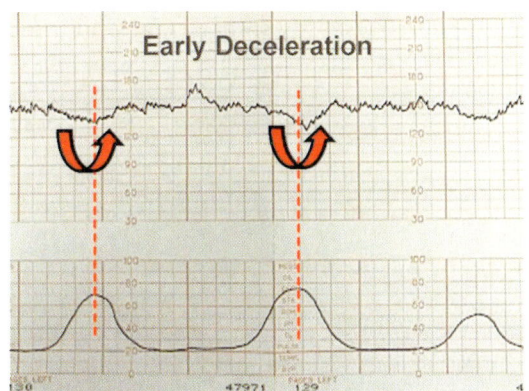

Fig. 14.4: CTG trace showing early decelerations

and do not require any intervention when associated with a good variability.[1]

Late Decelerations

Late decelerations are central nervous system reflex response to fetal hypoxia and acidemia (Fig. 14.5).[1]

Causes:
1. Maternal hypotension
2. Uterine tachysystole
3. Placental dysfunction
4. Placental abruption.

Late decelerations alone have a low predictive value for fetal hypoxia. Fetal heart rate variability and presence of accelerations are to be assessed. Normal FHR variability and

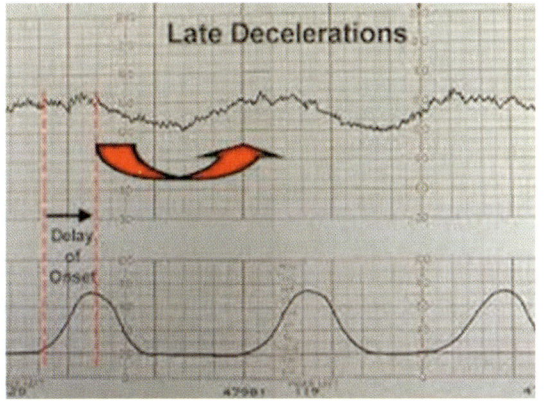

Fig. 14.5: CTG trace showing late deceleration.

presence of spontaneous or elicited accelerations are reassuring. Recurrent late decelerations with reduced variability and/ or absence of accelerations is ominous and indicative of fetal hypoxemia and prompt measures to improve fetal oxygenation and expidate delivery is indicated because persistence of this pattern is associated with increased fetal morbidity and mortality. Fetal blood sampling for pH or lactate level is indicated.

SINUSOIDAL PATTERN

A sinusoidal fetal FHR pattern is defined as a pattern of fixed, uniform fluctuations of the FHR that creates a pattern resembling successive geometric sine waves. It frequently is described as undulating and smooth and is characterized by the absence of variability.

"True" sinusoidal FHR patterns are associated with the following fetal conditions that result in either severe fetal anemia or severe/prolonged fetal hypoxia with acidosis.[1]

1. Chronic fetal anemia associated with erythroblastosis fetalis, usually from Rh sensitization
2. Acute, intrapartum asphyxia
3. Fetal-maternal hemorrhage
4. *In utero*, fetal hemorrhage.

When a "true" sinusoidal FHR pattern is noted, immediate intervention is required. Ultrasound evaluation for signs of fetal anemia (middle cerebral artery PSV), signs of hydrops, placental anomalies, fetal biophysical profile is indicated. Percutaneous umbilical blood sampling for fetal hemoglobin assessment and if needed intrauterine transfusion as indicated. Kleihauer–Betke test is indicated in suspected fetomaternal hemorrhage. If intrauterine resuscitation is not possible, preparations for emergency delivery should be initiated.

VARIABILITY ASSESSMENT AND MANAGEMENT

Saltatory Pattern

The saltatory FHR pattern is defined as fetal heart variability of >25 beats per minute.[1] This should alert the physician to look for and correct possible causes of acute hypoxia. Although it is a non-reassuring pattern, operative intervention for this finding in isolation is not warranted.

Reduced Variability

Reduced variability is defined as baseline FHR variability of less than 5 bpm[1] (Fig. 14.6).

Causes of reduced variability include

1. Fetal sleep
2. Maternal medications (magnesium sulphate, opioids)
3. Fetal hypoxia
4. Abnormalities of the fetal CNS
5. Extreme prematurity.

If the previous CTG trace has been normal with no decelerations, it is reasonable to assume that reduced variability is due to fetal sleep. One can also attempt to induce accelerations by scalp stimulation. Presence of

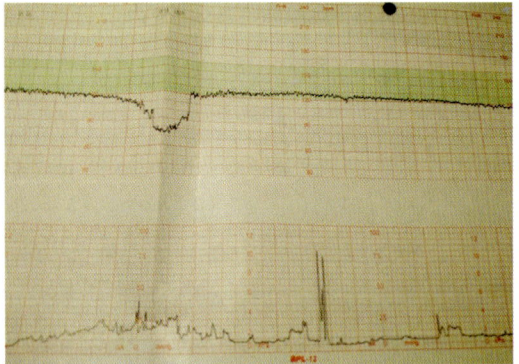

Fig. 14.6: 26-year-old primigravida at 37 weeks with fetal growth restriction and oligohydramnios was admitted with history of reduced fetal movements. On admission , the CTG trace showed reduced variability and a spontaneous deceleration. The trace did not improve with fetal stimulation. Emergency LSCS was performed to deliver a 1964 g baby. Liquor was meconium stained.

accelerations strongly rules out fetal hypoxia. Loss of variability with late decelerations is ominous and fetal scalp blood sampling for pH and lactate is indicated.

UTERINE ACTIVITY ASSESSMENT AND MANAGEMENT

Tachysystole

Tachysystole is defined as more than 5 contractions in 10 minutes.[1] Uterotonic drugs should be discontinued in the presence of tachysystole. Since uterine contractions causes intermittent interruption of blood flow to the intervillous space, excessive uterine activity that exceeds the critical level for an individual fetus will result in fetal hypoxemia. If there is no prompt response to stopping the uterotonic, a tocolytic drug may be used, such as terbutaline 250 mcg subcutaneously, as uterine relaxation may improve placental blood flow and fetal oxygenation.

Management Based on FIGO Guidelines

The FIGO consensus guidelines on intrapartum monitoring (2015) for interpretation of the CTG trace based on various parameters and their management[1] is described in Table 14.1.

Management of Suspicious Trace

Patients with suspicious tracings should be evaluated for factors that may reduce fetal oxygenation, taking into account associated clinical circumstances (e.g. abruption, trial of labour after a previous cesarean delivery, intrauterine growth restriction), and the stage and progress of labour.

Resuscitative measures can be initiated. Continued surveillance and frequent reassessment is indicated. Ancillary tests (described later) can be performed for more information. If suspicious trace that later becomes normal, no intervention is required. Patients with persistent suspicious trace are to be closely monitored for progression to pathological

Table 14.1: FIGO consensus guidelines on intrapartum fetal monitoring (2015)			
CTG classification cariteria, interpretation and recommended management			
	Normal	Suspicious	Pathological
Baseline	110–160 bpm	Lacking at least one characteristic of norma-	< 100 bpm
Variability	5–25 bpm	lity, but with no patho-logical features	Reduced variability for > 50 minutes, increased variability for > 30 minutes, or sinusoidal pattern for > 30 minutes
Decelerations	No repetitive decelerations		Repetitive* late or prolonged decelera-tions during > 30 minutes or 20 minutes if reduced variability, or one prolonged deceleration with > 5 minutes
Interpretation	Fetus with no hypoxia/acidosis	Fetus with a low probability of having hypoxia/acidosis	Fetus with a high probability of having hypoxia/acidosis
Clinical management	No intervention necessary	Action to correct reversible causes if identified, close monitoring or additional methods to evaluate fetal oxygenation	Immediate action to correct reversible causes, additional methods to evaluate fetal oxygenation, or if this is not possi-ble expedite delivery. In acute situa-tions (cord prolapse, uterine rupture or placental abruption) immediate delivery should be accomplished

The presence of accelerations denotes a fetus that does not have hypoxia/acidosis, but their absence during labour is of uncertain significance.

*Decelerations are repetitive in nature when they are associated with more than 50% of uterine contractions.

trace, which indicates fetal acidosis and where the delivery of the fetus is indicated.

Management of Pathological Trace

These findings are associated with an increased risk of fetal hypoxic acidemia, which can lead to cerebral palsy and neo-natal hypoxic ischemic encephalopathy. An effective and timely intervention before the development of severe acidosis can pre-vent neonatal morbidity and mortality (Fig. 14.6).

Resuscitative measures to improve utero-placental perfusion should be initiated simultaneously while making preparations for delivery. If there is no improvement in the FHR tracing after resuscitative measures, delivery should be expedited.

Algorithm for managing suspicious or pathological trace is depicted in Flow Chart 14.1.

In Utero Resuscitation

The following are the general measures for management of suspicious and pathological tracings which are aimed at improving uteroplacental perfusion:

1. Maternal repositioning onto her left or right side. Changing the maternal position to relieve the aortocaval compression and improve maternal blood flow to the placenta. It also sometimes relieves the cord compression as well (Fig. 14.7).

2. Administer oxygen. There is no evidence from randomized clinical trials that this intervention, when performed in isolation, is effective when maternal oxygenation is adequate.

3. Administer an intravenous (IV) fluid bolus (e.g. 500 to 1000 mL of lactated Ringer's or normal saline solution) for maternal hypotension.

Flow Chart 14.1: Algorithm for management of abnormal CTG.

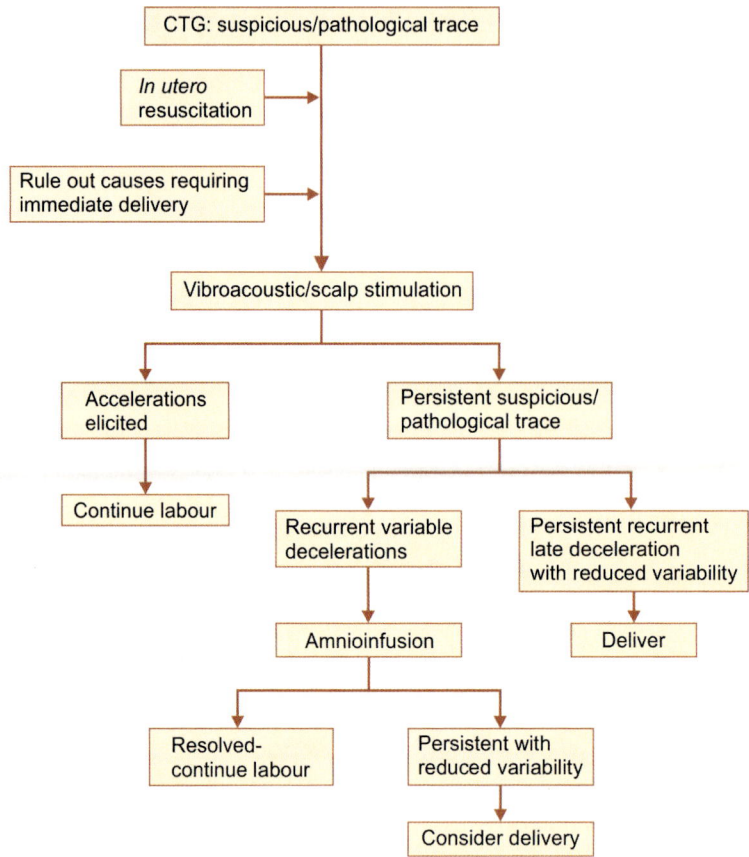

4. Discontinue uterotonic drugs (Fig. 14.8).
5. Amnioinfusion to reduce cord compression.
6. Administer a tocolytic drug (e.g. terbutaline 250 mcg subcutaneously).
7. Evaluate the patient for hypotension if patients is on epidural anesthesia administration of an alpha-adrenergic agonist like phenylephrine, ephedrine may reduce sympathetic blockade.

CTG Interpretation-Pitfalls

Conditions other than hypoxia can produce CTG tracing similar to suspicious or pathological patterns.

1. **Fetal sleep cycle:** It may last up to 40 minutes and is associated with decreased variability and reduced frequency of accelerations.

2. Inaccurate recording of the FHR pattern due to the faulty pick-up of the maternal iliac pulse. Simultaneous recording of the maternal pulse rate helps to distinguish whether the trace is fetal or maternal.

3. Technical factors like the speed of the CTG paper, e.g. at 3 cm/minute variability appears reduced to a clinician familiar with the 1 cm/minute scale.

4. Maternal intake of drugs, maternal fever and fetal conditions like cardiac arrhythmias and pre-existing neurological damage to the fetus can present with abnormal CTG tracing.

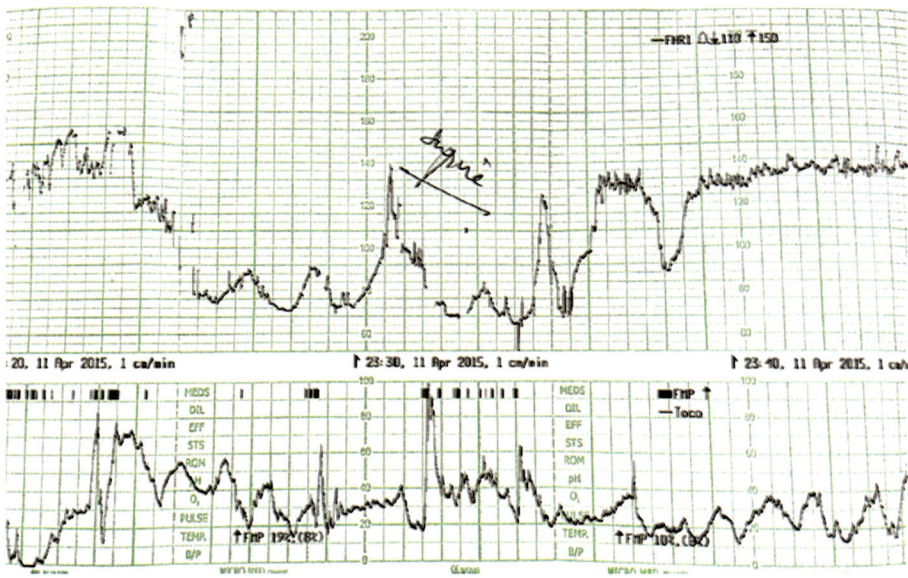

Fig. 14.7: 23-year-old primigravida of 39 weeks gestation with normal pregnancy is admitted to the labour ward in spontaneous labour. The CTG trace shows prolonged deceleration when the mother was supine with subsequent normalisation of trace after maternal change of position.

Comment: Intervention such as changing maternal position is a first step to improve fetal condition. In this case there was an improvement and the patient had a normal vaginal delivery.

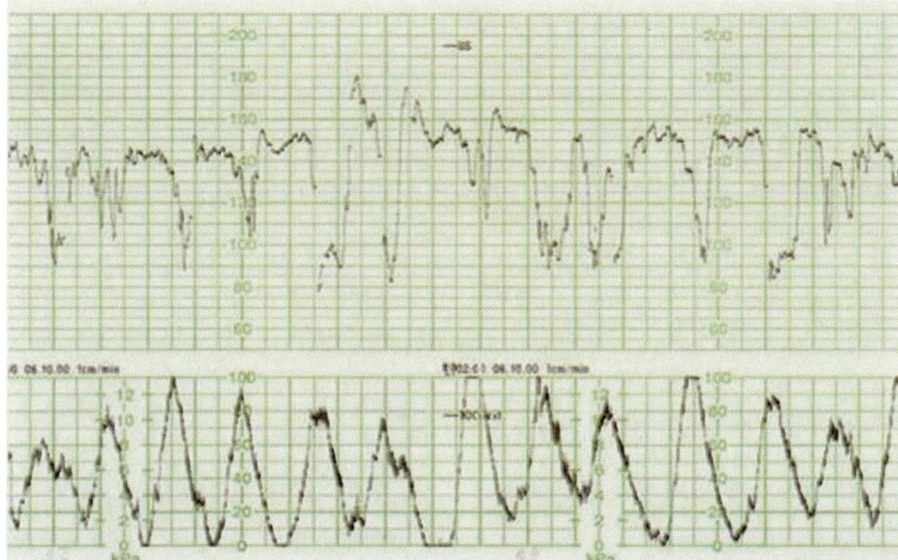

Fig. 14.8: 30-year-old primigravida, admitted with spontaneous rupture of membranes at 39 weeks of gestation. Clear liquor draining. Syntocinon infusion initiated for augmentation of labour. CTG trace later showed uterine tachysystole, 6 contractions/10 minutes with recurrent decelerations. Syntocinon infusion decreased due to over stimulation. Subsequently patient delivered vaginally.

CTG Need for Supplimentary Ancillary Tests

The positive predictive value of CTG for adverse outcome is low and the negative predictive value is high. While a normal CTG indicates reassuring fetal status, a non-reassuring or an abnormal CTG does not necessarily indicate 'fetal distress'.[1] Despite this, CTG is widely used for assessing fetal well-being in labour. To reduce such false positive cases and unnecessary medical interventions, additional adjunctive tests for assessment of the fetal well-being may be considered, which are complimentary to CTG. Several adjunctive technologies have been developed over the last decades, including Fetal Blood Sampling (FBS) for pH and lactate monitoring, fetal stimulation, pulse oximetry, and ST waveform analysis.[2]

Vibroacoustic Stimulation/Fetal Scalp Stimulation

Fetal scalp stimulation involves stimulating the fetal scalp by rubbing it with the examiner's fingers or using forceps to clasp the fetal skin, or alternatively using vibroacoustic stimulation applied to the mother's abdomen. The main purpose of this is to evaluate fetuses showing reduced variability on CTG to distinguish between deep sleep and hypoxia/acidosis. It is of questionable value in other patterns. When fetal stimulation leads to the appearance of an acceleration (acceleration of 15 bpm lasting for 15 seconds) and subsequent normalization of the fetal heart pattern, this should be regarded as a reassuring feature. However failure to provoke acceleration is not uniformly predictive of fetal acidemia.[2,3]

Fetal Scalp Blood Sampling

Fetal blood sampling during labour was first described in 1962 and is currently used for assessment of fetal blood gases and/or lactate. Fetal blood sampling may be used in cases of suspicious or pathological CTG. When pathological CTGs indicate a severe and acute event, immediate action should be taken and blood sampling is not advised as it would cause further delay. A low pH of less than 7.20[4,5] or a high lactate of more than 4.8 mmol/L[5] is considered abnormal. Fetal blood sampling is invasive and can be performed only if the cervix is at least 3 cm dilated and the membranes have ruptured. It is contraindicated in cases of chorioamniotinitis and suspected fetal coagulopathy.[2]

Fetal Pulse Oximetry

Though promising, fetal pulse oximetry has not been widely used. It directly measures the proportion of hemoglobin that is carrying oxygen. The authors of the Cochrane review in 2014 of seven trials involving 8013 women, concluded that addition of fetal pulse oximetry does not reduce the overall cesarean section rate. They concluded that a better method than pulse oximetry is required to enhance the overall well-being of the fetus in labour.[6]

STAN (ST Aanalysis)

ST analysis (STAN) involves a combination of FHR interpretation (concurrent CTG) and analysis of fetal electrocardiogram. A normal ST waveform is horizontal or upward sloping and a normal T wave has a constant amplitude. Both the ST segment and T wave are analysed by the software. Similar to ECG of adults with cardiac ischemia, cardiac hypoxia and ischemia in the fetus causes ST depression and inversion of T wave which is detected by the STAN software. It analyses the average waveform of fetal ECG signal over 30 consecutive heart beats and then compares this to the average of the subsequent 30 complexes. It is able to determine the ST segment and T wave changes over time and alerts the occurrence of 'STAN event'.[2] The authors of the Cochrane review in 2015 concluded that the STAN analysis had a modest reduction in operative deliveries and fetal blood sampling.[7] Their findings provided modest support for the use of fetal STAN only if CTG has non-reassuring features. If the CTG is normal, 'ST event' should be ignored. This can be used only after the rupture of membranes and is contra-indicated in suspected chorioamnionitis. This cannot be used if the ECG morphology is already abnormal at the beginning of the trace.

Practice Points

- CTG analysis needs to be integrated with other clinical information for a comprehensive interpretation and adequate management
- Good fetal heart rate variability is predictive of good fetal oxygenation
- Evaluate the overall picture: whether it is stable or deteriorating
- Look at the fetal reserve (post-dated, FGR, oligo-hydramnios)
- Consider delivery if no reassuring features (reduced/absent variability, no accelerations even after fetal stimulation)

REFERENCES

1. Ayres-de-Campos D, Spong CY, Chandraharan E. The FIGO Intrapartum Fetal Monitoring Expert Consensus Panel. FIGO consensus guidelines on intrapartum fetal monitoring: Cardiotocography. Int J Gynecol Obstet 2015;131(1):13–24.
2. Visser GH, Ayres-de-Campos D. FIGO Intra-partum Fetal Monitoring Expert Consensus Panel. FIGO consensus guidelines on intrapartum fetal monitoring: Adjunctive technologies. Int J Gynecol Obstet 2015;131:25–9.
3. Skupski DW, Rosenberg CR, Eglinton GS. Intrapartum fetal stimulation tests: a meta-analysis. Obstet Gynecol 2002;99(1):129–34.
4. Bretscher J, Saling E. pH values in the human fetus during labor. Am J Obstet Gynecol 1967; 97(7):906–11.
5. Wiberg-Itzel E, Lipponer C, Norman M, Herbst A, Prebensen D, Hansson A, et al. Determination of pH or lactate in fetal scalp blood in management of intrapartum fetal dis-tress: randomised controlled multicentre trial. BMJ 2008;336(7656): 1284–7.
6. East, Christine E, et al. Fetal pulse oximetry for assessment in labour. The Cochrane Library (2014).
7. Neilson J. Fetal electrocardiogram for fetal moni-toring during labour. The Cochrane Library (2015).

Role of the CTG Today

Rachael Yates

Fetal surveillance primarily includes cardiotocography (CTG) or Intermittent Auscultation (IA).[1]

Intermittent auscultation is the most appropriate form of fetal assessment for low-risk fetuses.[1]

Cardiotocography monitoring is for at risk or high-risk fetus.

As clinicians, we need to ensure this is performed correctly and the right choice of fetal surveillance is chosen. The practice of fetal surveillance during labour would be expected to detect those fetuses at risk of compromise, allowing appropriate intervention and thereby increasing the likelihood of improved perinatal outcomes.[1,3]

A huge benefit, of both doctors and midwives learning the same surveillance protocol is that all obstetric clinicians (doctors and midwives) are using the same professional language when describing a CTG.

Consistent professional training is essential in using the CTG today, whether it is the RANZCOG Fetal Surveillance Education Programme (FSEP) or the K2 Perinatal Training Programme. Whatever the program chosen, all clinicians need to understand the definitions and be speaking the same language.

RANZCOG FSEP and K2 Perinatal training programme are two standardised CTG learning programmes that are recommended for obstetricians and midwives. They have been developed using the best available evidence.[1] RANZCOG FSEP has an online learning component and face to face workshop followed by an assessment. The K2 Perinatal Training Programme is based on the UK Guidelines for fetal surveillance. The K2 Fetal Monitoring Training System is an interactive computer-based training system that can be accessed over the internet, anywhere, anytime.[1,2]

Any one programme can be used; however the importance of continual education in all aspects of midwifery and obstetrics has long been recognised all over the world, like Australia. Routine training has been difficult to implement, but I am pleased to say for the last few years, all doctors and midwives in South Australia, complete RANZCOG learnings for fetal surveillance.

The aim of the standardised Clinical Guideline is, in combination with continuing education and training of obstetric staff, to reduce adverse perinatal outcomes related to inappropriate or inadequate performance and/or interpretation of intrapartum fetal surveillance. This is achieved by encouraging best practice in:

- The decision to the use and interpret IA or continuous CTG
- Appropriate antenatal and perinatal risks identified and managed for each pregnant woman

- Management of suspected fetal compromise both pre labour and intrapartum
- Evidence to avoid routine admission CTG's.[1]

Monitoring the health of the fetus during labour has therefore become a key component of modern maternity care[1,3] (Tables 15.1 to 15.3).

Traditionally, this was undertaken by simple regular auscultation of the fetal heart with a stethoscope, pinnard and now fetal Doppler. However, this approach was considered by many to be inadequate, particularly for high-risk pregnancies. Therefore, in an effort to reduce the incidence of intrapartum fetal mortality and morbidity, the use of intrapartum electronic fetal monitoring (EFM), particularly continuous CTG, has steadily increased over the last 35 years.[1] Fetal surveillance in labour, whether by intermittent auscultation or by EFM, should be discussed with and recommended to all women antenatally.[1,3]

An admission CTG increases the rate of continuous EFM use and may increase the rate of cesarean section. The clinician needs to assess individual women's circumstances, risk to decide whether or not to use an admission CTG.[1] Regardless of the method of intrapartum monitoring, it is essential that an accurate record of fetal wellbeing is obtained. Fetal and maternal heart rates should be differentiated whatever the mode of monitoring used.[1]

The use of CTG for intrapartum fetal surveillance has now become embedded in practice without robust Randomised Controlled Trial (RCT) evidence to support it. The RCTs of continuous CTG which have been undertaken have suggested that its use is not associated with statistically significant improvements in long-term neonatal outcomes such as cerebral palsy, but that it is associated with significantly increased rates of (unnecessary) operative delivery. Nonetheless, not surprisingly, concerns about maternal hazards and small or absent perinatal benefit have led some authorities to advise against the routine use of continuous CTG for low-risk labours.[1]

Table 15.1: Antenatal indications for a CTG	
Risk factor	**Potential physiologiacal basis**
Decreased fetal movements	Reduced fetal reserves
Maternal hypertension/pre-eclampsia	Reduced placental function
Antepartum hemorrhage	Reduced placental functional area
Exogenous prostaglandins	Uterine stimulation
Clinical IUGR	Reduced fetal reserves
Diabetes requiring medication	Poor placentation or placental vascularisation
PPROM	Incareased risk of intrauterine infection
Threatened preterm labour	Reduced available reserves
Abdominal trauma	Placental shearing
Post dates pregnancy	Declining placental function
Maternal medical condition	Dependant on the condition, i.e. SLE
Spunous labour (long latent phase)	Reduced fetal reserves
Rhesus isoimmunization	Reduced fetal hemoglobin/O_2 carrying apacity
Known fetal abnormality	Reduced fetal capacity to tolerate contractions
Oligohydramnios	Reduced placental function and fetal reserves
Abnormal Doppler studies	Reduced placental perfusion

Table 15.2: Antenatal indicator for CTG monitoring in labour

Risk factor	Potential physiologiacal basis
Abnormal Doppler flows	Reduced placental perfusion (↑ resistance)
Antepartum hemorrhage	Reduced placental functional area
Abnormal antenatal CTG	Existing fetal compromise
Suspected or continued IUGR	Reduced fetal reserves
Oligohydramnios (or polyhydramnios)	Reduced placental function/↓ fetal reserves/↑ cord compression
Prolonged pregnancy ≥ 42 weeks	Declining placental function/↑ fetal morbidity and mortality
Multiple pregnancy	Increased fetal risk
Breech presentation	Increased fetal risk
Prolonged ROM > 24 hours prior to the onset of labour	Increased risk of intrauterine infection
Known fetal abnormality requiring monitoring	Reduced fetal capacity to tolerate with labour
Maternal hypertension or pre-eclampsia	Poor placental development and function
Prior uterine scar	Increased risk of uterine rupture
Fetal movements reduced (within the week preceding labour)	Possible pre-existing compromise
Diabetes requiring medication, poorly controlled or macrosomia	Poor placental vascularisation/macrosomia/↑ morbidity
Maternal medical conditions consulting a fetal risk	Dependant on the condition, i.e. SLE
Maternal obesity (MI ≥ 40)	Increased fetal risk
Maternal age ≥ 42	Increased fetal risk
Abnormal maternal screening associated with fetal compromise, i.e. low PAPP -A < 0.4 MOM	Possible pre-existing fetal compromise

http:www.fsep.edu.au/

This chapter will not go into detail, however, knowing the definition of a normal CTG is very important in using the CTG Today.

A **normal** CTG is associated with a low probability of fetal compromise and has the following features:
- Baseline rate 110–160 bpm.
- Baseline variability of 6–25 bpm.
- Accelerations of 15 bpm for 15 seconds.
- No decelerations.

All other CTGs are by this definition **abnormal** and require further evaluation taking into account the full clinical picture.[1,3]

An Example of the Normal CTG[3] (Fig. 15.1)

Trace description:
- With a baseline heart rate of 140 bpm
- Normal baseline variability 6–25 bpm

- Plentiful accelerations. (reactivity) and no decelerations.

Management:
- This trace would be classified as 'normal'
- No management is required. The CTG may be removed.

An Example of an Abnormal CTG[3] (Fig. 15.2)

Trace description
- The baseline FHR is 125 bpm
- Baseline variability is normal
- There are no accelerations recorded despite plentiful fetal movement being recorded
- There are no decelerations
- There is no uterine activity recorded.

Interpretation: The absence of accelerations is attributed to the high dose beta blockers, which suppress sympathetic innervation. If

Table 15.3: CTG monitoring in labour indicators	
Risk factor	**Potential physiological basis**
Abnormal auscultation or CTG	Inability to assess compromise by auscultation/ protential fetal compromise
Induction of labour with prostaglandin/oxytocin	Uterine hyperstimulation/fetal asphyxia
Oxytocin augmentation	Uterine hyperstimulation/fetal asphyxia
Regional anesthesia, i.e. epidural or spinal, paracervical block (pre-epidural CTG should be considered)	Maternal hypotension, fetal hypoxia, CTG changes
Abnormal vaginal bleeding in labour	Reduced placental function
Maternal pyrexia >38°C	Incraeased fetal O_2 requirements
Meconium or blood stained liquor	Prior fetal compromise
Absent liquor following amniotomy	Reduced placental function/↑ cord compression/ ↓ fetal reserves
Prolonged first stage, defined by referral guidelines	Reducing fetal reserves
Prolonged second stage, fefined by referral guidelines	Reduced fetal reserves, ↑ maternal and fetal acidosis
Preterm labour <37 weeks	Limited reserves
Tachysystole, i.e. >5 contractions over 10 minutes without FHR abnormalities	Risk of fetal asphyxia
Uterine hypertonus, i.e. contractions >2 minutes or within 60 seconds of each other, without FHR abnormalities	Risk of fetal asphyxia
Uterine hyperstimulation, i.e. tachysystole or uterine hypertonus with FHR abnormalities	Fetal asphyxia

http:www.fsep.edu.au/

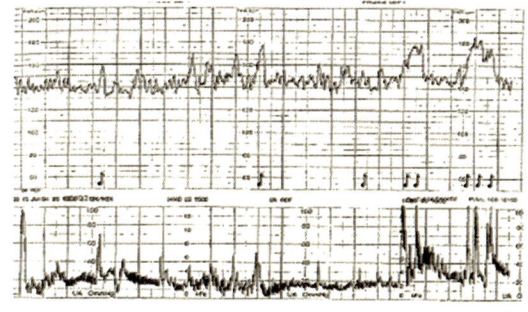

Fig. 15.1: A primigravida woman at term who reported reduced fetal movements.

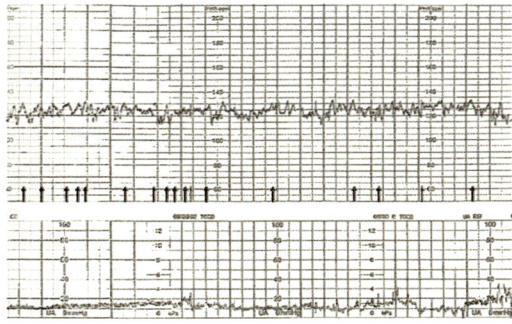

Fig. 15.2: A primigravida woman at 34 weeks gestation on high dose beta blockers for pregnancy induced hypertension.

fetal movement can be palpated, fetal well-being is assured.

Management/Action: Repeating the CTG immediately prior to the mother taking the next dose of medication may give the best chance of assessing accelerations, drugs, such

as beta blockers, are the second most common cause of reduced baseline variability and/or accelerations.

Included in Table 15.4, is a Fetal CTG Review Guide that is based on the RANZCOG

Table 15.4: CHSALHN fetal CTG review guide (Dr S Scroggs 2016)

CTG	The baseline fetal heart rate is assessed in the absence of fetal movement, uterine activity, acelerations and decelerations			
	Normal CTG	CTG requiring further evaluation		
Baseline rate (bpm)	110–160	100–109	>160 or rising baseline	<100 bpm for >5 minutes
Baseline variability	6–25		Reduced 3–5	Absent <3
Accelerations	Absent			
Decelerations	Nil	Early decelerations Variable decelerations	Complicated variables late prolonged	Complicated variables or late decelerations with reduced variability
Other abnormal features	Nil			Sinusoidal
Accelerations	2 present in 20 minutes (antenatal only)	<2 present in 20 minutes (antenatal only)		

(The absence of accelerations intrapartum is not considered abnormal)

Normal	Continue CTG throughout labour. As per SAPPG for fetal surveillance, Good Practice Note: 15/60 review documented on CTG and in medical record.
Seek senior midwifery advice	There is a low probability of fetal compromise. Discuss with more senior clinician to decide on plan. Document plan in Partograph. Fax CTG image to identified network if needed.
Discuss with medical staff	These features may be associated with significant fetal compromise and require further action. Immediate call to doctor to either present to hospital or fax to medical officer.
Expedite delivery and/or provide intra-uterine resuscitation	These featurs are likely to be associated with significant fetal compromise and require immediate management, which *may include* urgent delivery; Be aware of staff call in processes for urgent or Cat 1 LSCS, additional support or resources.

Guidelines. It has been developed for Country Health SA midwives and obstetric staff. It is a guideline to recognising and seeking review of a fetus that may be compromised. This attached to each CTG in Country Health SA.

CONCLUSION

The role of the CTG has an important place in obstetric clinical practice today. There are some considerations for best practice:

- Understanding the role of the CTG, when to use and the definitions of normal and abnormal
- Following a clinical practice guideline
- Education for all obstetric staff is a best

practice consideration. The education needs to be consistent so all obstetric clinicians are using the same language
- Fetal surveillance education is imperative for excellence in clinical CTG use and interpretation.

REFERENCES

1. https://www.fsep.edu.au/FSEP/media/FSEP IFS% 20Clinical%20Guideline/RANZCOG% 20IFS%20Clinical%20Guideline%203rd%20ed.% 202014.pdf.

2. http://www.k2ms.com/products/perinatal-training-programme.aspx

3. https://www.fsep.edu.au/Home.

Limitations of CTG Monitoring

Muralidhar V Pai, Vidyashree Poojary

CTG is a useful tool to detect intrapartum hypoxia for timely intervention to avoid its consequences. Intrapartum hypoxia and subsequent metabolic acidosis is associated with complications such as Hypoxic Ischemic Encephalopathy (HIE) and neonatal death or long-term complications such as cerebral palsy and learning difficulties.[1] While CTG has effectively picked up compromised fetuses, the incidence of cerebral palsy has not changed much whereas the incidence of instrumental delivery and cesarean sections have increased significantly.[2] Cardiotocography has its own limitations, and it is important for us to understand them for correct interpretation and action.

Pattern Recognition

Important aspect of reading a CTG is recognizing the abnormal patterns. The complexity of FHR patterns makes standardization difficult. There is no consistency in the recognition of these patterns among observers. The disagreements are usually in identifying different classes of decelerations, evaluation of variability, and categorization of tracings as suspicious and pathological.[3,4] Previous studies suggest that both the inter- and intra-observer variability of FHR is inadequate, with clinicians interpreting the FHR similarly in less than one-third of the cases, and the same clinicians reaching a different conclusion two months later when reviewing the same tracing.[5] Maternal bearing down efforts in second stage makes CTG less reliable.[6]

High Sensitivity but Very Poor Specificity and Positive Predictive Value

CTG has a high sensitivity but very poor specificity and positive predictive value in predicting intrapartum fetal hypoxia/acidosis and false positive rate is very high.[7,8] This may lead to unnecessary operative intervention without any substantial benefit to perinatal outcome.[4,9] Hence CTG cannot substitute for good clinical observation and judgement, during labour. All said and done the specificity of CTG for prediction of cerebral palsy is low even in the presence of multiple late decelerations or decreased variability.

Maternal Artifact during Intrapartum Fetal Heart Rate Monitoring

CTG can pick-up maternal heart rate, which may mimic FHR patterns and result in Maternal Heart Rate Artifact (MHRA). Risk factors for MHRA include an active fetus, twin gestation, and obesity. A prospective study done by Stephanie et al. in 2014 showed a higher incidence of MHRA and in more than 2% of women in labour it may lead to adverse outcomes.[10]

Requirement for Additional Tests of Fetal Well-being

Computerized analysis of the CTG seems to be better than subjective interpretation by the individuals. However, it is better to do additional tests such as Fetal scalp Blood Sampling (FBS), fetal scalp lactate, fetal pulse oximetry, fetal ECG also called STAN or ST analyzer in view of poor positive predictive value of CTG for metabolic acidosis and it has high false positive rate. These other tests will be dealt in details in the other Chapters in this book.

Knowledge of Adverse Neonatal Outcome Alters Clinicians' Interpretation of the Intrapartum Cardiotocograph

The knowledge of adverse neonatal outcome leads to a more severe classification of the intrapartum CTG, as revealed by a study conducted in 2011, in which the obstetricians were asked to analyze the CTG tracings according to FIGO guidelines before and after giving the information of neonatal outcome.[11]

Randomized controlled trials comparing continuous CTG monitoring with intermittent auscultation were conducted between 1970 and 1990s, using different CTG interpretation criteria, so it is difficult to establish their validity in current clinical practice. Implementation of conclusions of these studies showed only a 50% reduction in neonatal seizures but no differences in the incidences of overall perinatal mortality and cerebral palsy.[12]

Clinical guidelines should be simple and objective, so as to allow rapid decision-making even in complex and stressful situations. Regular and structured training of all stakeholders is essential to ensure proper use of this technology.

REFERENCES

1. CEMACH—Centre for Maternal and Child Enquiries (previously Confidential Enquiries into Maternal and Child Health). Perinatal Mortality Surveillance 2004: England, Wales and Northern Ireland. London: CEMACH, 2006.
2. Alfirevic Z, Devane D, Gyte GM. Continuous cardiotocography (CTG) as a form of electronic fetal monitoring (EFM) for fetal assessment during labour. Cochrane Database Syst Rev 2006;3:CD006066.
3. Blackwell SC, Grobman WA, Antoniewicz L, Hutchinson M, Gyamfi-Bannerman C. Inter-observer and intraobserver reliability of the NICH 3-tier fetal heart rate interpretation system. Am J Obstet Gynecol 2011;205(4):378.e1–5.
4. Diogo Ayres-de-Campos, Catherine Y. Spong, Edwin Chandraharan. FIGO consensus guidelines on intrapartum fetal monitoring: Cardiotocography. Int J Gynecol and Obstet 2015;131:13–24.
5. Chauhan SP, Klauser CK, Woodring TC, Sanderson M, Magann EF, Morrison JC. Intrapartum nonreassuring fetal heart rate tracing and prediction of adverse outcomes: interobserver variability. Am J Obstet Gynecol 2008;199:623.e1–623.e5.
6. Tullio G, Giovanni M, Federica B, Paola R, Francesca G, Nicola R, Tiziana F, Gianluigi P. Cardiotocographic findings in the second stage of labor among fetuses delivered with acidemia: a comparison of two classification systems. European J Obstet & Gynecol and Rep Biol 203 (2016) 297–302.
7. FIGO consensus guidelines on intrapartum fetal monitoring.
8. Alfirevic Z, Devane D, Gyte GML. Continuous cardiotocography (CTG) as a form of electronic fetal monitoring (EFM) for fetal assessment during labour. Cochrane Database of Systematic Reviews 2013; 5: CD006066.
9. O'Mahony F, Hofmeyr GJ, Menon V. Choice of instruments for assisted vaginal delivery. Cochrane Database Syst Rev 2010;11:CD005455.
10. Stephanie P, Felipe M, Kelli R, Zachary M. Ferraro, Lawrence O. The Incidence of Maternal Artefact during Intrapartum Fetal Heart Rate Monitoring. J Obstet Gynaecol Can 2014;36(11): 962–8.
11. Ayres-de-Campos D, Arteiro D, Costa-Santos C, Bernardes J. Knowledge of adverse neonatal outcome alters clinicians' interpretation of the intrapartum cardiotocograph. BJOG 2011;118: 978–84.

17 Analysing Other Methods of Fetal Monitoring

Amanda Henry, Daniella Susic

ANALYSING OTHER (NON-CTG) METHODS OF FETAL MONITORING

This chapter aims to illustrate forms of fetal monitoring that are used in addition to CTG, and to explore their uses, benefits and limitations.

The most important starting point is determining the risk of the individual pregnancy and flag which women require increased antenatal surveillance. This process is complex and vital to plan appropriate antenatal care.

There are several methods which can be used to perform antenatal fetal monitoring. A discussion about fetal movements, fundal height, ultrasound assessment including; estimated fetal weight, amniotic fluid volume, Doppler measurements and biophysical profile and an outline of the management of a small for gestational age fetus follows.

These various methods of fetal monitoring can be used alone, or in combination, to establish fetal growth and well-being depending on the resources available in the location providing the antenatal care. The primary goal of antenatal fetal monitoring is to ensure that timely intervention can be achieved to prevent poor obstetric outcomes and stillbirth.

Whilst antenatal fetal monitoring has been a key part of obstetric clinical practice since the 1970s its role has not been proven in large randomised clinical trials. Its efficacy has been based either on: (1) observational studies reporting lower rates of fetal death in monitored pregnancies, or (2) equal or lower rates of fetal death in monitored high-risk pregnancies compared to an unmonitored low-risk obstetric population. The one component of fetal monitoring that has been rigorously evaluated with randomised trials are Doppler-based tests.

DETERMINING RISK

In order to establish which pregnancies need increased fetal monitoring it is important to determine the baseline risk of each pregnancy. In particular, what factors are present that may affect the growth and well-being of the fetus during its development, and increase the risk of stillbirth or other poor pregnancy outcome. A broad (but by no means exhaustive) list of indications for fetal surveillance are shown in Table 17.1. Most of the listed factors increase the risk of Fetal Growth Restriction (FGR), which is associated with approximately six times increased risk of term stillbirth, and is not well-detected by most methods of fetal monitoring.[1]

ASSESSMENT TECHNIQUES

Fetal Movements

Maternal perception of fetal movements is often the first sign of fetal presence or well-being a woman experiences. Conversely, a

Table 17.1: Factors which increase the risk of Fetal Growth Restriction (FGR)	
Demographic and social factors	Advanced maternal age
	Obesity*
	Drug use: Tobacco
	In vitro fertilisation
Past medical history	Diabetes: Pregestational, type 1 and type 2*
	Hypertensive disorders
	Systemic lupus erythematosus
	Antiphospholipid syndrome
	Sickle cell disease
Past obstetric history	Prior fetal demise
	Prior FGR
Maternal disease	Alloimmunisation
	Gestational diabetes requiring treatment, or poorly controlled on diet alone*
	Pre-eclampsia
Fetal concerns	Clinically suspected fetal growth restriction
	Major fetal structural anomalies
Pregnancy complications	Twin pregnancy or higher order multiples
	Postdates pregnancy
	Decreased fetal activity
	Oligohydramnios or polyhydramnios
	Abnormalities in first and second trimester
	Down Syndrome maternal analyte screening results
	Major placenta previa

* although obesity and diabetes are typically associated with macrosomia, they may also be associated with FGR

decrease or change in perceived fetal movements is often the first sign of fetal compromise. The diagnosis of Decreased Fetal Movements (DFM) needs to be taken seriously, and prompt assessment to determine fetal well-being arranged. Whilst 40% of pregnant women become concerned about decreased fetal movements at some point, mostly this is transient. However, 4–15% of women will have sufficient concern to contact their pregnancy care provider.[2]

Maternal perception of DFM flags pregnancies with increased risk of adverse outcome. A Norwegian prospective, population-based registry of 2313 women with a singleton pregnancy in the third trimester showed that although many women had fetal death at the time of their DFM presentation, perinatal mortality rate amongst women presenting with DFM and a live fetus was much higher than the general obstetric population: 8.2/1000 versus 2.9/1000.[2]

Twenty-two percent of third trimester DFM have been associated with poor outcomes, including impaired fetal growth, preterm birth, poor neonatal condition, and emergency delivery.[3] Fetal growth restriction appears to be a major factor contributing to the increased risk of adverse neonatal outcome in pregnancies with DFM.

Early recognition and investigation of DFM, through ultrasound as well as CTG, may therefore provide an opportunity to identify compromised fetuses and prevent adverse outcome, through increased surveillance or delivery.

Fundal Height

As FGR is associated with adverse pregnancy outcome, the first line instrument to establish adequate fetal growth is the bedside symphyseal-fundal height measurement (Fig. 17.1). The accuracy of the fundal height measure-

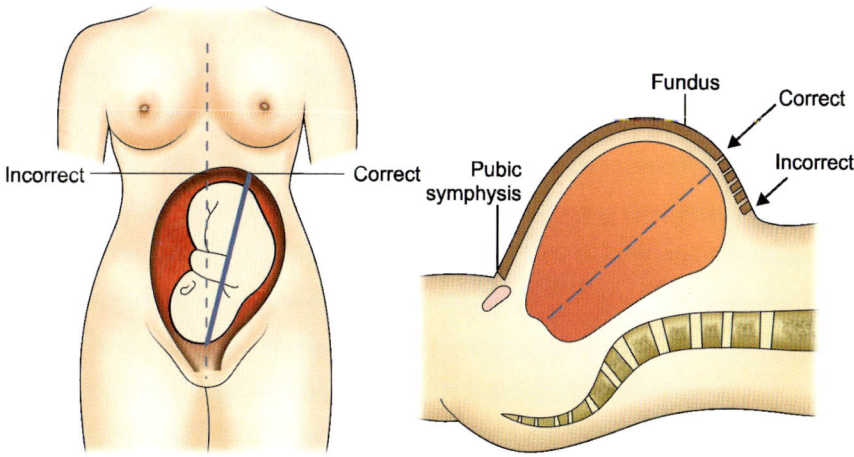

Fig. 17.1: Correct angle and anatomical landmarks to measure fundal height.

ment is increased, and bias reduced, if the same practitioner performs the measurements from the top of the symphysis to the uterine fundus using the unmarked side of the tape measure (Fig. 17.2).[4] Whilst the symphysis-fundal height is a screening method to detect fetal growth outside of normal parameters (prompting ultrasonographic evaluation of

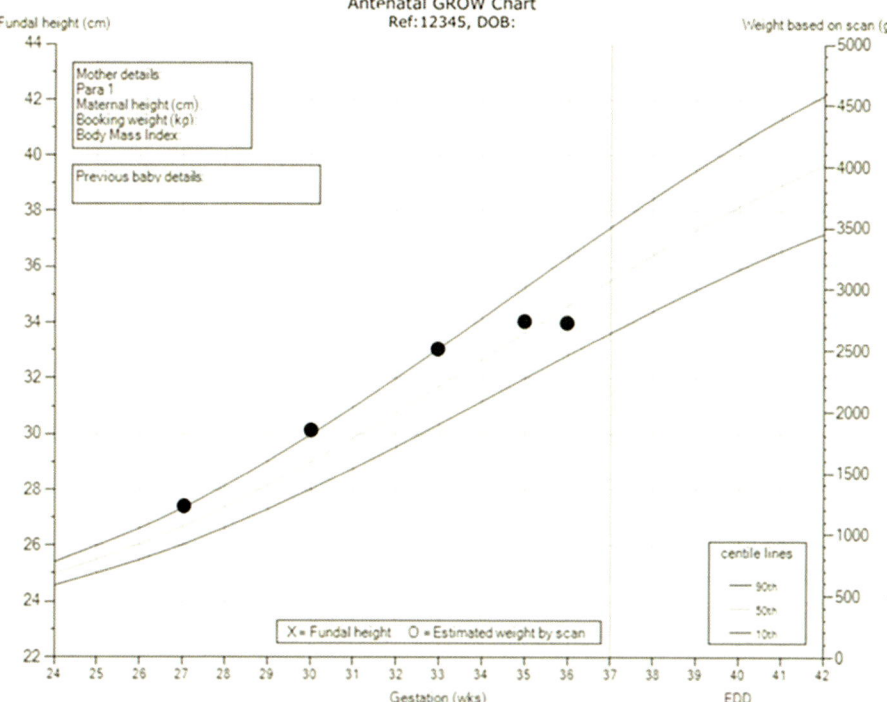

Fig. 17.2: An example of a customised growth chart showing plateauing fundal height still within a measurement range that would not trigger further evaluation using standard fundal height measurement practice.

fetal growth if the measurement is >2 cm above or below expected gestational age), it is important to acknowledge factors that will affect its sensitivity; increased maternal body mass index, bladder volume, parity, uterine fibroids and ethnic group.[5]

Whilst not used widely, there is now also evidence that using customised fundal height charts doubles the antenatal detection of small for gestational age fetuses in nulliparous women.[6] Gardosi reported that the detection of small for gestational age fetuses increased from 29.2 to 47.9% with the utilisation of customised fundal height charts.[1]

Ultrasonographic Evaluation

The evaluation of ultrasound needs to be discussed using the individual components in order to fully appreciate its benefits and limitations in the use of antenatal monitoring.

Fetal Growth

Once a clinical assessment of suspected FGR or abnormal fetal growth is made, further ultrasound evaluation is recommended.

Estimated Fetal Weight (EFW)

Fetal weight estimation using ultrasound is derived from a number of different mathematical formulas. Equations incorporating abdominal circumference, biparietal diameter and femur length appear to provide the most accurate estimation of fetal weight.[7] EFW has an accepted inaccuracy and weight value ranging from 15% either side of the calculated EFW. In population studies, EFW only identifies approximately one-third of third-trimester small-for-gestational-age (SGA) fetuses (<10th centile), although in experienced centres also using Doppler approximately 60% are identified.[8] The accuracy is best when there is severe FGR <3rd percentile. If being used to predict actual birth weight, it performs best when measured within 1–2 weeks of birth.[9]

Abdominal Circumference

Ultrasound measurement of fetal Abdominal Circumference (AC) reflects abdominal adipose tissue and hepatic size, both of which decrease in the malnourished growth restricted fetus. Most studies report AC as the most sensitive single morphological marker of FGR.[9] AC measurement alone is not helpful in evaluating an asymmetrically growth restricted fetus. However, the AC measurement was validated in a study of 3616 pregnancies over 25 weeks of gestation that had a single ultrasound examination performed within two weeks of delivery, predicting SGA infants. It was also more predictive for fetal growth restriction than measurement of either head circumference of biparietal diameter, or the combination of AC with either one of these two variables. The optimal time to screen for fetal growth restriction was 34 weeks gestation,[9] although other studies suggest 36 weeks.[8]

Amniotic Fluid Volume

Amniotic Fluid Volume (AFV) reflects fetal well-being, as the hypoxic fetus diverts blood to vital organs and away from the kidneys. This leads to decreased renal perfusion and urine production, resulting in reduced AFV over time.

Ultrasound examination is the only practical clinical method of assessing amniotic fluid volume. It is performed routinely in every second and third trimester ultrasound examination. There are many causes for an abnormal AFV apart from FGR to consider, including premature rupture of membranes and fetal congenital anomalies. Abnormal AFV is associated with an increased risk of a variety of perinatal outcomes, hence detecting an abnormal AFV should prompt further fetal assessment.

Semi-qualitative methods: A study of 45 pregnancies comparing SDP with direct measurement of amniotic fluid at cesarean delivery

Table 17.2: Interpretation of results between the Single Deepest Pocket (SDP), or Miximal Vertical Pocket (MVP) and the Amniotic Fluid Index (AFI)			
Method	**Oligohydramnios**	**Normal**	**Polyhydramnios**
SDP	≤ 2 cm	≥ 2 cm <8 cm	≥ 8 cm
AFI	≤ 5 cm	>5 cm and <24 cm*	≥ 24 cm

*An AFI result of 5.1 to 8 would be considered borderline in some institutions, but the clinical implications of this are unclear

identified only 18% of the pregnancies with low amniotic fluid volume.[10] A 2008 systematic review of randomised trials found the use of AFI increased the rate of diagnosis of oligohydramnios [RR 2.3], induction of labour [RR 2.1], and caesarean delivery for fetal distress [RR 1.5] without improving peripartum outcomes (Table 17.2).[11]

Ultrasound assessment of amniotic fluid volume can also be made qualitatively by an experienced sonographer scanning the entire uterine contents and reporting: oligohydramnios, normal liquor volume, or polyhydramnios. This method results in similar sensitivity when compared with semiqualitative methods, and is therefore just as valid a screening tool given increased intervention without improvement of outcome when using AFI.

Doppler Velocimetry

The use of fetal and maternal blood vessel flow measurements provides information about uteroplacental blood flow, and fetal responses to changing physiology. In response to fetal acidosis, there is a stepwise progression of changes (most clearly seen in preterm severe FGR). Umbilical artery Doppler indices increase when 60 to 70% of the placental vascular tree is compromised.[12] The next stage involves middle cerebral artery impedance falling and fetal aortic resistance increasing to preferentially direct blood to the fetal brain and heart. Once this occurs there is ultimately cessation or reversal of end diastolic flow in the umbilical artery, with increased fetal system resistance evident in the ductus venosus and inferior vena cava.[13]

Umbilical Artery (UA)

UA Doppler assessments are most useful for monitoring fetuses with early-onset growth restriction due to uteroplacental insufficiency. A cascade from a normal high flow, low resistance system in the diastolic cardiac phase of a healthy fetus to a diminished, absent or reversed diastolic flow in a growth restricted fetus is associated with worsening destruction of placental villus vasculature (Figs 17.3A to D). This cascade of fetal compromise to the point of absent or reversed end diastolic flow is reflective of fetal hypoxia and academia, and increased perinatal morbidity and mortality.[14]

In a systematic review of 16 randomised trials including over 10,000 high-risk pregnancies, the use of Doppler ultrasound resulted in a variable decrease in perinatal mortality (OR 0.71, 95% CI 0.52–0.98, 1.2 versus 1.7%, number needed to treat 203).[15] A further systematic review of five randomised trials inclusive of 14,000 low-risk pregnancies found routine umbilical artery screening did not improve perinatal outcomes.[16]

Middle Cerebral Artery

The Middle Cerebral Artery (MCA) peak systolic velocity is best used to monitor pregnancies at risk of fetal anemia such as alloimmunisation. The MCA Doppler measurement used to calculate the cerebroplacental ratio is emerging as a predictor of adverse outcome for both appropriately grown and growth restricted fetuses.[18] Although high-level evidence does not yet

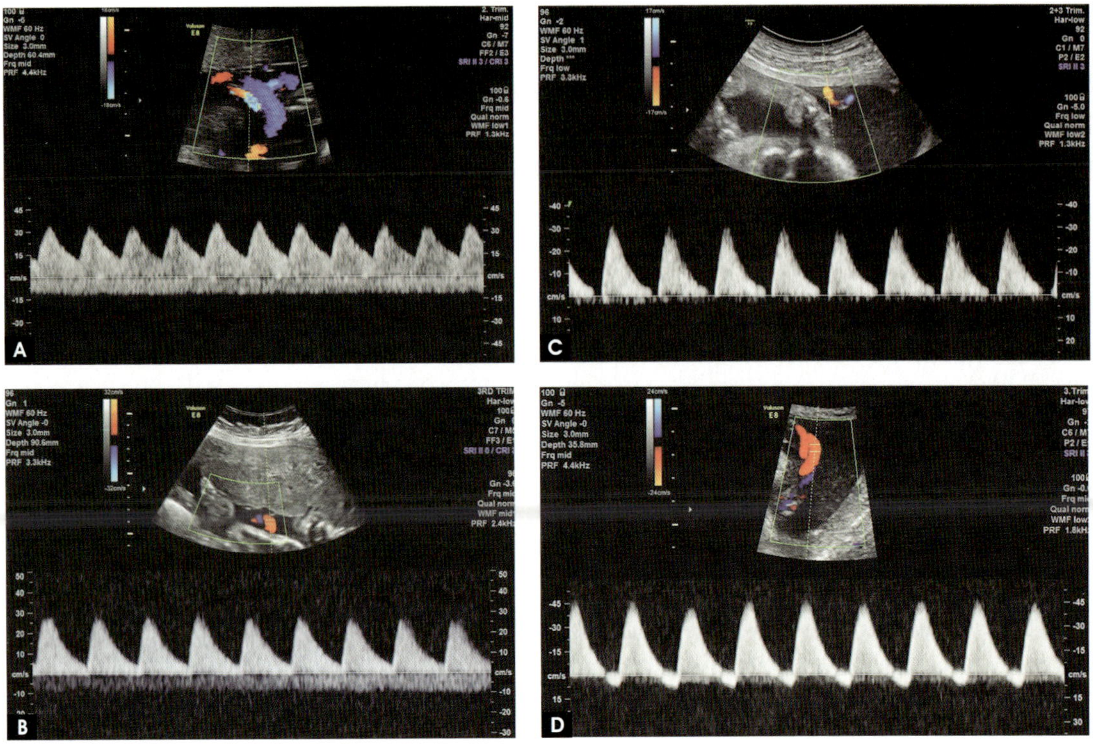

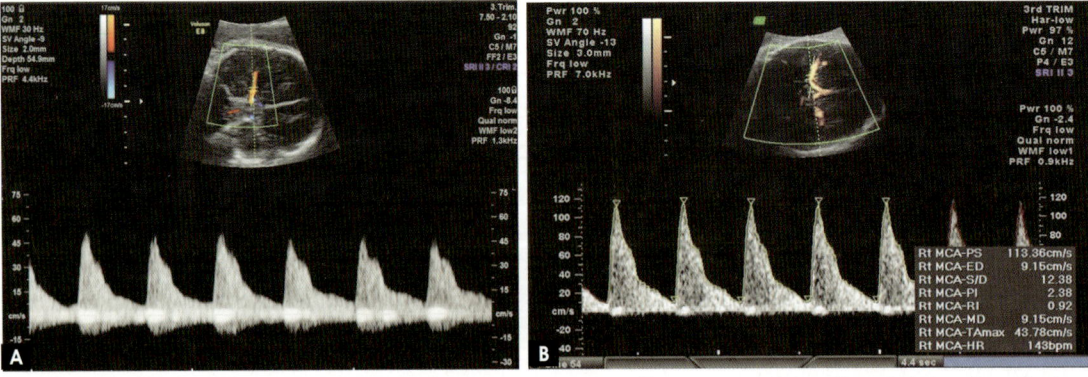

Figs 17.3A to D: Doppler flow patterns of clinical importance to Umbilical Artery (UA) waveform assessment.[17] (A) Normal UA flow –end-diastolic flow clearly seen well above the spectral baseline, (B) Increased resistance– Minimal end-diastolic flow is present above the spectral baseline. Doppler indices for gestation (31 weeks) would be used to confirm that increased resistance for gestation is present, (C) Absent end-diastolic flow–No flow seen end-diastole, (D) Reversed end-diastolic flow–flow drops below spectral baseline in end-diastole, indicating reversal of flow.

support its routine use, it is likely to be most useful in assessing late preterm/early term fetuses, where UA Doppler is usually in the normal range even in FGR, but MCA Doppler shows evidence of brain sparing (Figs 17.4A to C).

Figs 17.4A and B: Middle Cerebral Artery (MCA) Doppler waveforms.[17](A) Normal MCA pattern 3rd trimester with sharp systolic upstroke, then low diastolic velocity, end-diastolic flow preserved, (B) High peak systolic velocity in anemic (31 week) fetus: note change of scale (PRF 7.0 vs. 4.4 kHz in 4A, and change in spectral Y-axis scale) in order to display the full MCA waveform in this case.

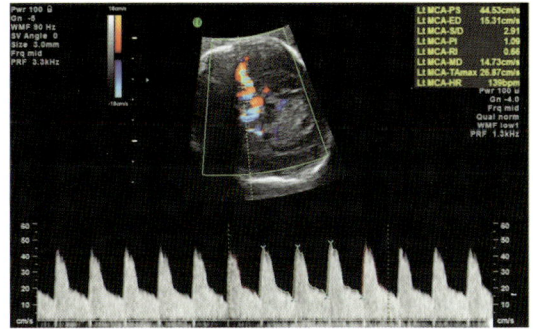

Fig. 17.4C: Middle Cerebral Artery (MCA) Doppler wave-forms.[17]—MCA waveform from fetus at 29 weeks with IUGR (estimated weight <3rd centile and absent umbilical artery end-diastolic flow): High diastolic velocities, low PI and RI.

Fetal Veins

Fetal venous Doppler evaluation in the umbilical vein, ductus venosus and inferior vena cava is best used when looking at conditions that affect cardiovascular function, or in severe early-onset placental insufficiency. These measurements are technically challenging and are best utilised by experienced sonographers assessing severe growth restriction, twin-twin trans-fusion syndrome, fetal hydrops and fetal arrhythmias.

Flow Chart 17.1: Assist in determining which fetus need increased monitoring and what methods are best chosen (adapted from)[20]

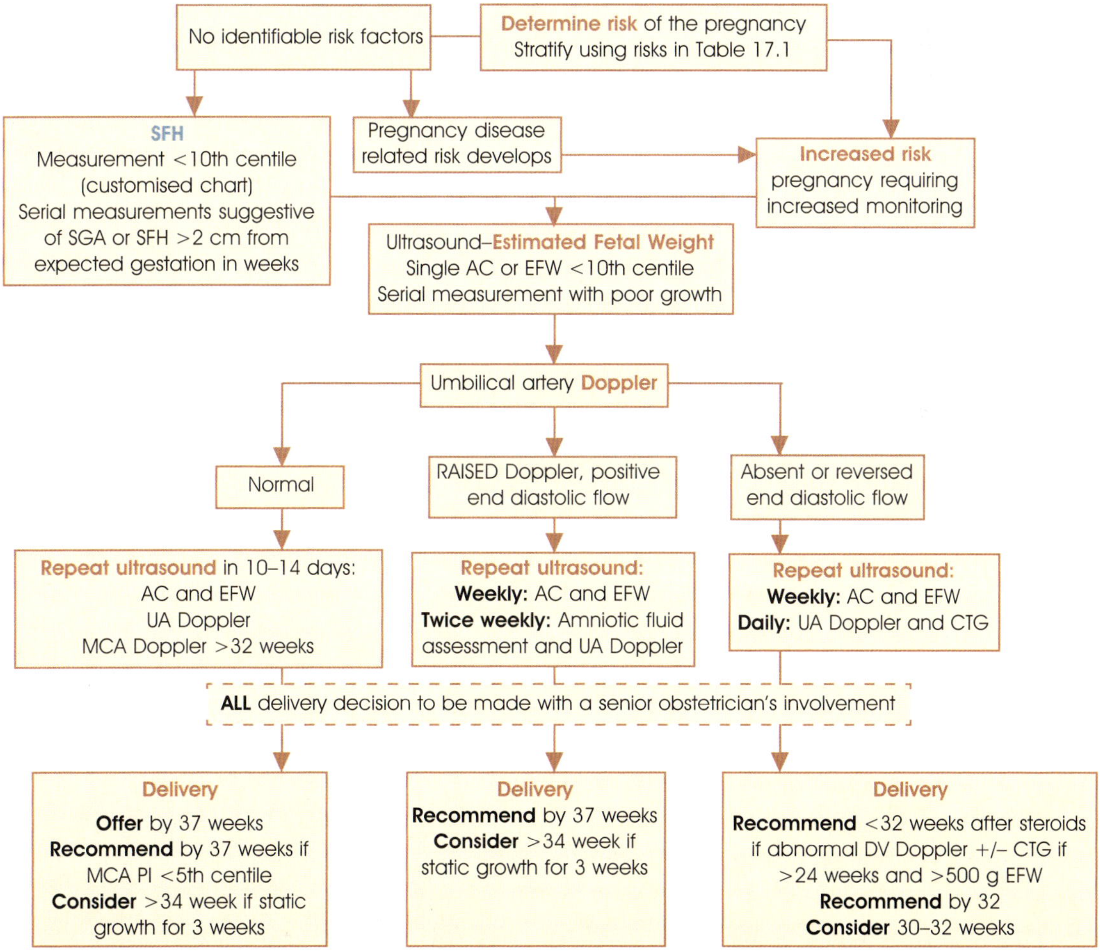

```
                    No identifiable risk factors          Determine risk of the pregnancy
                                                          Stratify using risks in Table 17.1

        SFH                              Pregnancy disease                        Increased risk
Measurement <10th centile               related risk develops                pregnancy requiring
(customised chart)                                                           increased monitoring
Serial measurements suggestive
of SGA or SFH >2 cm from        Ultrasound–Estimated Fetal Weight
expected gestation in weeks     Single AC or EFW <10th centile
                                Serial measurement with poor growth

                            Umbilical artery Doppler

        Normal              RAISED Doppler, positive        Absent or reversed
                            end diastolic flow              end diastolic flow

Repeat ultrasound in 10–14 days:    Repeat ultrasound:          Repeat ultrasound:
AC and EFW                           Weekly: AC and EFW           Weekly: AC and EFW
UA Doppler                           Twice weekly: Amniotic fluid Daily: UA Doppler and CTG
MCA Doppler >32 weeks                assessment and UA Doppler

        ALL delivery decision to be made with a senior obstetrician's involvement

        Delivery                         Delivery                         Delivery
Offer by 37 weeks              Recommend by 37 weeks          Recommend <32 weeks after steroids
Recommend by 37 weeks if        Consider >34 week if            if abnormal DV Doppler +/– CTG if
MCA PI <5th centile            static growth for 3 weeks           >24 weeks and >500 g EFW
Consider >34 week if static                                          Recommend by 32
growth for 3 weeks                                                   Consider 30–32 weeks
```

Keys: SFH=Symphyseal-Fundal Height, SGA=Small for Gestational Age, AC=Abdominal Circumference, EFW=Estimated Fetal Weight, UA=Umbilical Artery, MCA=Middle Cerebral Artery, PI=Pulsatility Index, DV=Ductus Venosus, CTG=Cardiotocograph

Biophysical Profile

The Fetal Biophysical Profile (BPP) combines a CTG with an ultrasound assessment of fetal well-being. It involves, in addition to presence or absence of reactive CTG, a 30 minute ultrasound assessment of presence or absence of: ≥ 1 episode of fetal breathing movement lasting ≥ 30 seconds, ≥ 3 discrete body or limb movements, ≥ 1 episode of extremity extension with return to flexion or opening or closing of a hand, maximum vertical amniotic fluid pocket of ≥ 2 cm. Each component is assigned a score of two points if criteria are met or zero points if not. The maximum score is 10/10, a normal result is $\geq 8/10$ or 8/8 without CTG, equivocal 6/10 and abnormal is ≤ 4. The BPP has a good false negative rate of 0.07–0.08%, but high false positive rate of 40–50% so it is unwise to make a delivery decision on this parameter alone.[19]

Choosing which method of assessment: Careful evaluation of baseline maternal risk, gestational age, test availability and cost needs to be considered. The Flow Chart 17.1 provides one example of which tests may be used for surveillance and to determine appropriate timing of delivery when there is evidence of fetal compromise shown by EFW < 10th centile and/or abnormal fetal Doppler studies.

REFERENCES

1. Gardosi J, Madurasinghe V, Williams M, et al. Maternal and fetal risk factors for stillbirth: population based study. BMJ 2013; 346: f108.
2. Tveit JV, Saastad E, Bordahl PE, et al. The epidemiology of decreased fetal movements. Proceedings of the Norwegian Perinatal Society Conference, November 2006.
3. Dutton PJ, Warrander LK, Roberts SA, et al. Predictors of poor perinatal outcome following maternal perception of reduced fetal movements– a prospective cohort study. PloS One 2012;7: e39784.
4. Jelks A, Cifuentes R, Ross MG. Clinician bias in fundal height measurement. Obstet Gynecol 2007:110:892–9.
5. Papageorghiou AT, Ohuma EO, Gravett MG, et al. International standards for symphysis-fundal height based on serial measurements from the Fetal Growth Longitudinal Study of the INTER-GROWTH-21st Project: prospective cohort study in eight countries. BMJ 2016;335:i5662.
6. Roex A, Nikpoor P, Eerd E, et al. Serial plotting on customised fundal height charts results in a doubling of the antenatal detection of small for gestational age fetuses in nulliparous women. ANZJOG 2012;52:78–82.
7. Anderson NG, Joley IJ, Wells JE. Sonographic estimation of fetal weight: comparison of bias, precision and consistency using 12 different formulae. Ultrasound in Obset Gynecol 2007;30: 173–9.
8. Sovio U, White IR, Dacey A, Pasupathy D, Smith GC. Screening for fetal growth restriction with universal third trimester ultrasonography in nulliparous women in the Pregnancy Outcome Prediction (POP) study: a prospective cohort study. Lancet 2015;386:2089–97.
9. Warsof SL, Cooper DJ, Little D, Campbell S. Routine ultrasound screening for antenatal detection of intrauterine growth retardation. Obstet Gynecol 1986;67:33–9.
10. Horsager R, Nathan L, Leveno KJ. Correlation of measured amniotic fluid volume and sonographic predictions of oligohydramnios. Obstet Gynecol 1994;83:955–8.
11. Nabhan AF, Abdelmoula YA. Amniotic fluid index versus single deepest vertical pocket as a screening test for preventing adverse pregnancy outcome. Cochrane Database Syst Rev 2008; CD006593.
12. Thompson RS, Trudinger BJ. Doppler waveform pulsatility index and resistance, pressure and flow of the umbilical placental circulation: an investigation using a mathematical model. Ultrasound Med Biol 1990;16:449–58.
13. Ferrazzi E, Bozzo M, Rigano S, et al. Temporal sequence of abnormal Doppler changes in the peripheral and central circulatory systems of the severely growth-restricted fetus. Ultrasound Obstet Gynecol 2002;19:140–6.
14. Karsdorp VH, van Vugt JM, van Geijn HP, et al. Clinical significance of absent or reversed end diastolic velocity waveforms in umbilical artery. Lancet 1994;344:1664–8.
15. Alfirevic Z, Stampalija T, Gyte GM. Fetal and umbilical Doppler ultrasound in high-risk pregnancies. Cochrane Database Syst Rev 2013: CD007529.

16. Alfirevic Z, Stampalija T, Medley N. Fetal and umbilical Doppler ultrasound in normal pregnancy. Cochrane Database Syst Rev 2015; CD001450.

17. Henry A. The use of the fetal myocardial performance index in complicated pregnancies. PhD thesis 2016; UNSW, Sydney.

18. DeVore GR. The importance of the cerebro-placental ratio in the evaluation of fetal well-being in SGA and AGA fetuses. Am J Obstet Gynecol 2015;213:5–15.

19. Dayal AK, Manning FA, Berck DJ, et al. Fetal death after normal biophysical profile score: An eighteen-year experience. Am J Obstet Gynecol 1999;181:1231–6.

20. Royal College of Obstetricians and Gynaecologists. Small-for-Gestational-Age Fetus, investigation and management (Green-Top Guideline Number 31, 2nd ed). 2013, RCOG, United Kingdom. Accessed via: https://www.rcog.org.uk/en/guidelines-research-services/guidelines/gtg31/August 25, 2017.

Clinical Scenarios in High-Risk Pregnancies

Nuzhat Aziz, Sridevi Veluganti Nagasai

Cardiotocography (CTG) is the recommended method for antenatal surveillance and intrapartum fetal monitoring in high-risk women. A high-risk pregnancy can be defined as any pregnancy where the probability of adverse outcome is greater than that of an uncomplicated pregnancy, or a reference population. The responsible factors can be constitutional or pre-existing medical conditions, or a pregnancy related complication developing in the antenatal or intrapartum period.[1,2]

Antenatal Surveillance in High-Risk Pregnancy

Antenatal fetal surveillance with CTG is done with an assumption that chronic asphyxia follows a specific pattern and we may be able identify fetal compromise with CTG and ultrasound. These tests follows history, physical examination, lab tests and ultrasound based fetal growth surveillance. Acute events of fetal compromise cannot be identified with CTG surveillance, such as abruption, rupture uterus, etc. The aim of antepartum surveillance is to optimise timing of delivery and to deliver before the fetus develops irreversible hypoxic damage.[3]

Fetal Adaptations to Hypoxia

The fetus is programmed to live in a low oxygen environment. The major adaptive mechanisms are fetal hemoglobin's higher ability to extract oxygen, the shunting of blood through ductus venosus and a higher maternal cardiac output; all facilitating better oxygen delivery to fetus. Chronic hypoxia induces adaptive changes; fetal hemoglobin rises, the cardiac output increases with centralization of blood flow to vital organs (brain sparing effect). Severe and continued hypoxic insult leads to a stage of cardiac dysfunction, loss of cerebral adaptation leading to metabolic acidosis, reflected by loss of baseline variability in CTG (Fig. 18.1).

Maternal hypoxia can lead to fetal hypoxia resulting in abnormal CTG patterns, which improve on correction of maternal condition. The interpretation of an antenatal trace should take into consideration all four fetal heart rate characteristics; baseline rate, baseline variability, accelerations and decelerations. Nonstress test is labelled as reactive and nonreactive based on the presence of fetal accelerations in a specific time period. Loss of baseline variability is the most important predictor of chronic hypoxia. Variability is largely influenced by the sympathetic and the parasympathetic interactions and the gestational age. Sympathetic system develops earlier in contrast to the parasympathetic system's maturation in the third trimester. Hence CTG in the second trimester and early third trimester may not show good baseline variability. Studies on chronic hypoxia have shown the sequence of deterioration to be umbilical artery Doppler changes, cerebral redistribution, reduced fetal breathing movements, decrease in amniotic fluid volume then reduced FHR baseline variability. Persistant hypoxia then leads to umbilical venous changes, loss of fetal movements and tone leading to bradycardia and fetal death.

The ideal time interval for antenatal fetal surveillance differs depending on the maternal and fetal condition, rate of disease progression. CTG should be used only when the fetus has crossed the period of viability and not before 26 weeks, when an intervention can be taken.

Fetal monitoring in critically ill mothers

Critical life-threatening events do occur in obstetrics. The fetus is dependent on the mother for its life—for oxygen through the uterine blood flow and placental circulation. The uterine circulation in a term pregnancy has a blood flow of 600–700 ml per minute, through the uterine arteries. Placental circulation is different with low resistance high blood flow with a strong alpha adrenergic system, which makes it extremely susceptible to endogenous or exogenous catecholamines. Maternal hemodynamic instability with shunting of blood to vital organs leads to a fall in uteroplacental circulation resulting in fetal compromise. Chronic decrease in the uterine artery blood flow is seen in maternal conditions like pre-eclampsia due to decrease in the number and narrowing of the spiral arteries. All conditions causing uteroplacental insufficiency and fetal growth restriction should be regarded as having a decreased uteroplacental reserve.[4]

The commonly observed acute causes of decreased uteroplacental circulation are maternal hypovolemia and aortocaval compression. The intervillous circulation is maximal in the lateral resting position and least during the uterine contractions. Maternal hypovolumia stimulates the protective vital organ sparing mechanism with constriction of peripheral vasculature and splanchnic

Table 18.1: Factors decreasing uteroplacental blood flow

1. Supine position: Aortocaval compression
2. Uterine contractions
3. Decreased in surface area of placenta
   a. Abruption
   b. Infarcts
   c. Thrombosis in fetal vessels
4. Maternal hypotension
5. Placental diseases
   a. Fibrin deposition
   b. Mesenchymal dysplasia

circulation. The uterine circulation is a part of the splanchnic circulation and may get affected by physiological response and by many pressor agents used during resuscitation. The conditions associated with decrease in uterine blood flow are shown in Table 18.1.

The maternal oxygen saturations are often affected by many factors in critically ill mothers. The placental blood flow and exchange are so designed that the fetal venous (umbilical vein–oxygenated fetal blood) saturations are almost similar to that of maternal venous blood. A fetus becomes acidotic when the aerobic metabolism is affected, which occurs if the fetal oxygen saturations fall below 30–35%, pO_2 falling below 15–20 mm Hg. This information is essential for fetal survival when there is maternal hypoxemia in respiratory diseases, for mothers on ventilatory support.

Maternal anemia in critically ill parturients results in decreased oxygen carrying capacity. The fetal oxygen delivery in a parturient with severe anemia on a ventilator can be improved remarkably by correction of anemia.[4]

Intrapartum Fetal Monitoring in High-Risk Women

All guidelines recommend continuous CTG monitoring in the intrapartum period of high-risk pregnancies.[5] The indications for use of electronic fetal monitoring are given in Table 18.2. Few more indications for CTG

Table 18.2: High-risk categorization[6]		
Maternal risk factors	**Fetal risk factors**	**Labour related risk factors**
Age <18 or >35 years	Multiple pregnancies	Suspicious FHR by IA
BMI > 35	Fetal growth restriction	Induced labour
Hypertension	Preterm labor	Augmented labour
Diabetes	Breech presentation	Prolonged labour
Cardiac disease	Rh isoimmunization	Epidural analgesia
Hemoglobinopathy	Oligohydramnios	Thick meconium
Severe anemia	Abnormal Dopplers	Vaginal bleeding
Hyperthyroidism	Previous stillbirth	Hyperstimulation
Collagen disease	Bad obstetric history	Previous CS
Renal disease	Intrauterine infection	
Hemodynamic instability for any reason	Post-term pregnancy	

Table 18.3: Additional indications for cardiotocographic monitoring[2]

- Maternal HR >120 per minute, on 2 occasions
- Temperature

Single reading of 100.4 °F

Two readings of 99.5 °F

- Continuous pain different from contractions
- Proteinuria of 2+
- Contractions that last for 60 seconds
- More than 5 contractions in 10 minutes

monitoring have been added by the National Institute on Clinical Excellence (NICE) through an update on intrapartum care guideline in June 2017 (Table 18.3).[2]

Examples of CTG Monitoring in High-Risk Pregnancies

Case 1

Primigravida, 24 years of age, BMI 29, admitted at 34 weeks gestation with eclampsia. Her BP at admission was 170/110 mm Hg, heart rate of 140/minute, respiratory rate of 34/minute, SpO$_2$ of 92% on room air, irritable responding to verbal commands. Brisk deep tendon reflexes with urine albumin of 4+. Obstetric examination showed a symphysio fundal height of 28 cm, cephalic presentation. Lab investigations showed platelets of 80,000/cumm, S Creatinine of 1.8 mg/dL, LDH of 880 IU/L, SGPT of 132 IU/L giving a diagnosis of eclampsia with HELLP syndrome (hemolysis, elevated liver enzymes, and low platelet count). Arterial blood gases revealed metabolic acidosis with a pH of 7.25, Base decifit of −8. Bedside CTG at admission is shown in Fig. 18.1.

Discussion: This pattern represents acute on chronic hypoxia. Ultrasound done showed severe fetal growth restriction with absent end diastolic flow in the umbilical artery, with oligohydramnios (AFI 4.5). The decelerations are a response to additional fetal stress due to uterine contractions. Delivery has to be performed after maternal stabilization and checking her coagulation profile with adequate blood products standby. Cesarean section was done and baby delivered; birth weight 1.2 kg, Apgar of 3, 4 and 6 with cord umbilical artery pH of 6.9, base deficit−16. Baby required ventilator support, had seizures on day 1 of life, recovered to be shifted to mothers side on Day 6. Long-term follow-up is essential for babies with chronic insults.

Case 2

Primigravida, 28 years of age was referred with severe pre-eclampsia at 32 weeks of gestation. She was complaining of headache

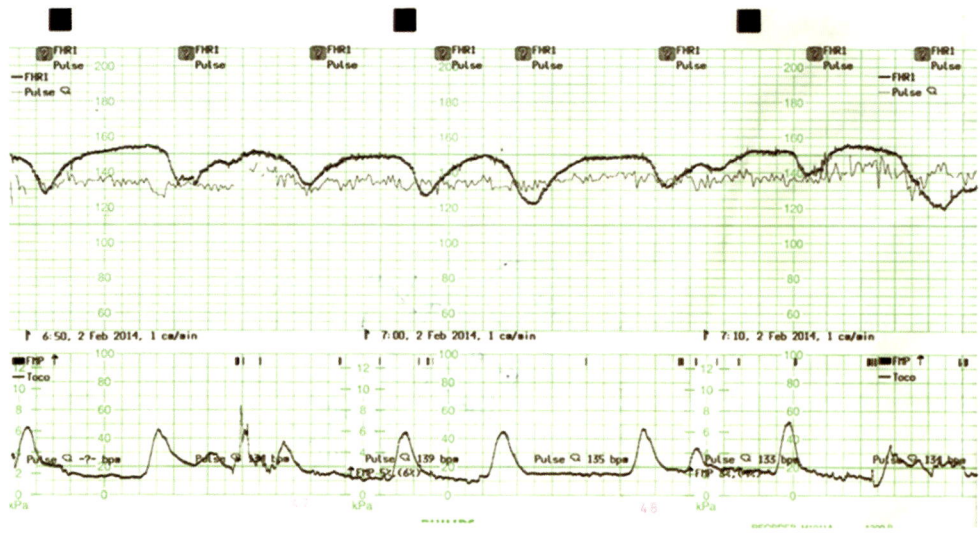

Fig 18.1: Bedside CTG at admission.

CTG Interpretation using the pneumonic DR C BRAVADO[7]		
DR	Define Risk	Metabolic acidosis, eclampsia, HELLP syndrome
C	Contractions	4 in 10 minutes
BRA	Baseline rate	150 beats per minute
V	Variability	Absent
A	Accelerations	None
D	Decelerations	Late decelerations
O	Opinion	Pathological trace, requiring delivery

and on examination her BP was 160/110 mm Hg with a heart rate of 100/minute, respiratory rate 25/minute, SpO$_2$ of 96% and brisk deep tendon reflexes. NST done at admission showed an on-reactive trace with decreased variability (Fig. 18.2). Ultrasound revealed fetal growth restriction, reversed end diastolic flow in umbilical artery and amniotic fluid index of 6.5.

Discussion: Decreased variability is the single most important marker for hypoxia. It was found to have the best correlation with metabolic acidosis at birth. Decreased variability with decelerations is an ominuos sign of impending death, and should stimulate a rapid response from the obstetric team.

Case 3

G4 A3, 28 years of age, BMI 32, with 35 weeks gestation, reporting to emergency room with pain abdomen of one hour duration. NST at admission showed a pathological trace (Fig. 18.3).

Discussion: Her previous three pregnancy losses were early trimester before 10 weeks, antiphospholipid antibody screen negative. Her pregnancy was uneventful with an ultrasound showing normal growth at 34 weeks. Abruptio placentae should be a part of differential diagnosis of any complaint of pain abdomen and preterm labour. Excessive uterine contractions of five or more

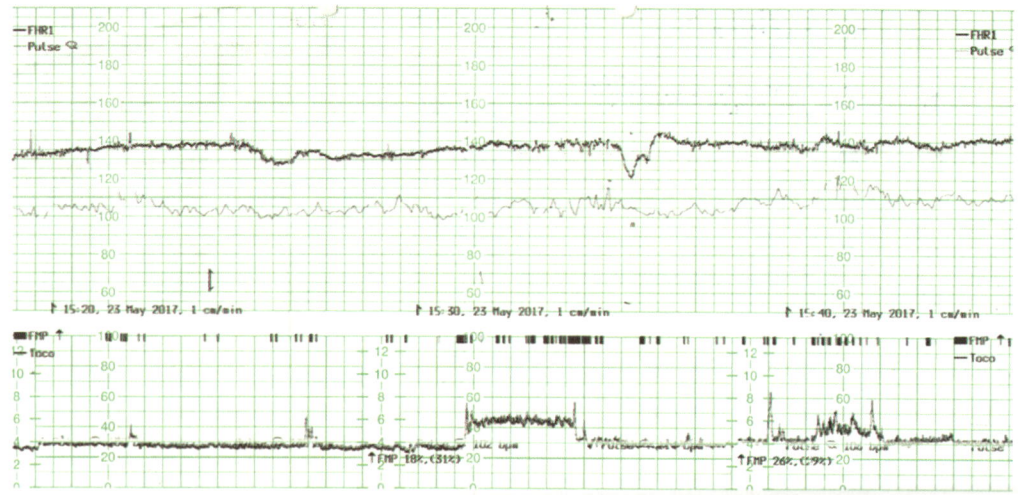

Fig. 18.2: Decreased baseline variability with normal baseline rate.

CTG Interpretation using the pneumonic DR C BRAVADO		
DR	Define Risk	Severe pre-eclampsia, preterm
C	Contractions	None
BRA	Baseline rate	140 beats per minute
V	Variability	Absent
A	Accelerations	None
D	Decelerations	None
O	Opinion	Pathological trace, requiring urgent delivery

contractions in 10 minutes are very common in induced or augmented labours. Tachysystole in spontaneous labour should stimulate an investigation for abruption (Fig. 18.4). The CTG changes are dependent on the amount of separation of the placenta. Here the decision to delivery interval was 15 minutes and a baby with Apgar of 4, 7 and 8 was delivered, birth weight of 2.2 kg, and cord arterial pH of 7.05, base excess of −10. A rapid response and delivery protocol helps in timely delivery in obstetric emergencies like cord prolapse, abruption, rupture uterus or scar dehiscence.

Case 4: Rupture uterus

Trial of labour after cesarean section for a G2 P1 L1 with 38 weeks pregnancy in active labour. She reached up to 6 cm cervical dilatation when CTG showed sudden onset severe prolonged deceleration (Fig. 18.5). The decision to delivery interval for cesarean section was 15 minutes and the baby was born 3.4 kg, with an Apgar of 3, 4 and 6 with an umbilical cord pH of 6.9, base deficit of −20. Rupture uterus with the fetus in the peritoneal cavity. The baby was discharged home on day 6 of life.

Case 5: Pseudosinusoidal Pattern and Sinusoidal Pattern

A Primigravida with term gestation, uneventful antenatal period is admitted in labour at 4 cm cervical dilatation. The cardiotocography done showed a trace as shown in Fig. 18.6. This pattern reverted back to normal in a short-time.

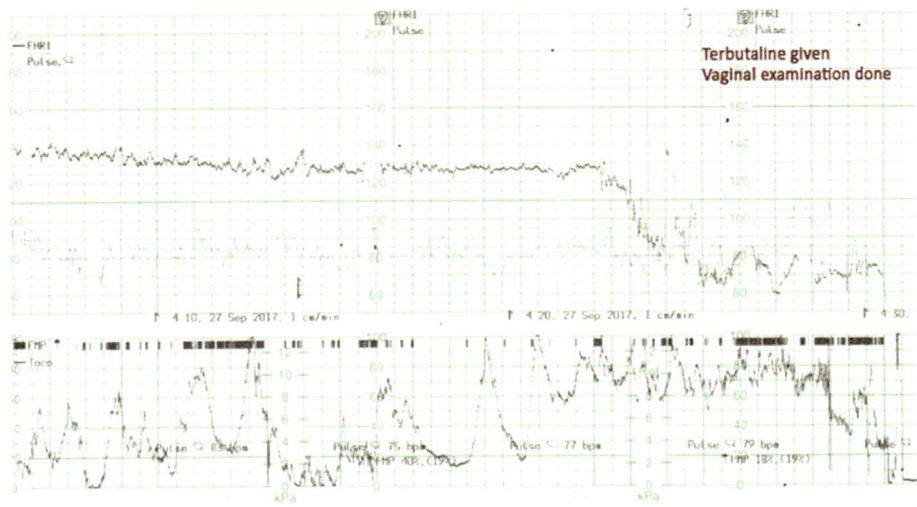

Fig 18.3: Abruption with ¾ separated placenta.

CTG Interpretation using the pneumonic DR C BRAVADO		
DR	Define Risk	Pain abdomen, preterm, recurrent pregnancy losses
C	Contractions	5 in 10 minutes
BRA	Baseline rate	130 beats per minute
V	Variability	Normal
A	Accelerations	None
D	Decelerations	Prolonged deceleration, not recovering with terbutaline
O	Opinion	Pathological trace, requiring urgent delivery.

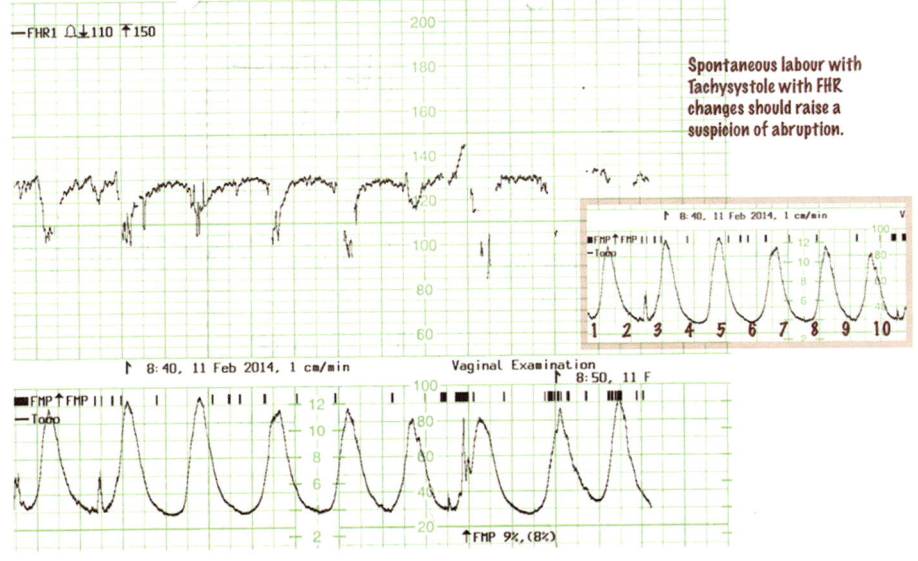

Fig. 18.4: Tchysystole with FHR change.

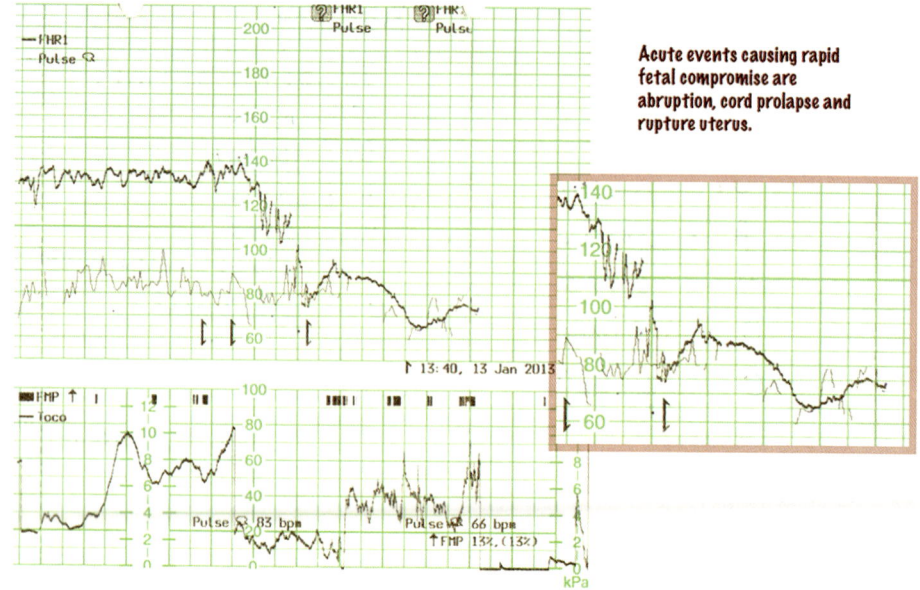

Acute events causing rapid fetal compromise are abruption, cord prolapse and rupture uterus.

Fig. 18.5: Rupture uterus.

CTG Interpretation using the pneumonic DR C BRAVADO		
DR	Define Risk	Pain abdomen, preterm, recurrent pregnancy losses
C	Contractions	Difficult to make out, 2 in 10 minutes
BRA	Baseline rate	135 beats per minute
V	Variability	Normal
A	Accelerations	None
D	Decelerations	Prolonged deceleration
O	Opinion	Pathological trace, requiring urgent delivery.

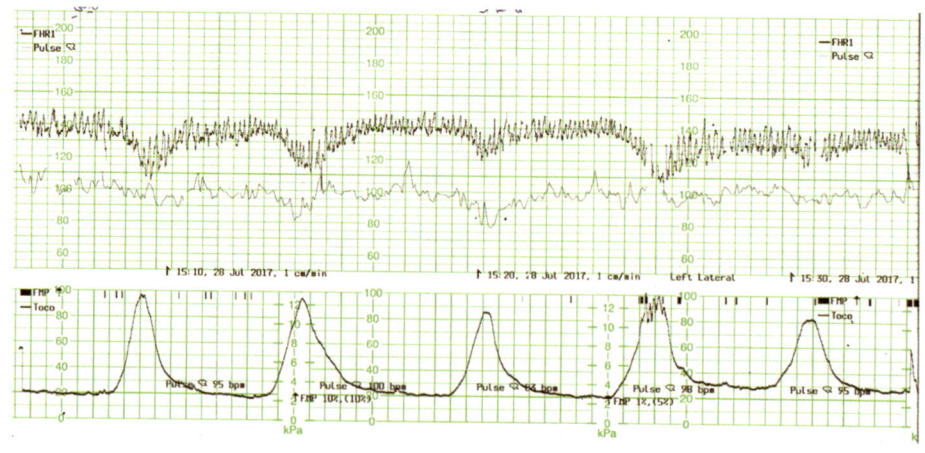

Fig. 18.6: Pseudosinusoidal pattern

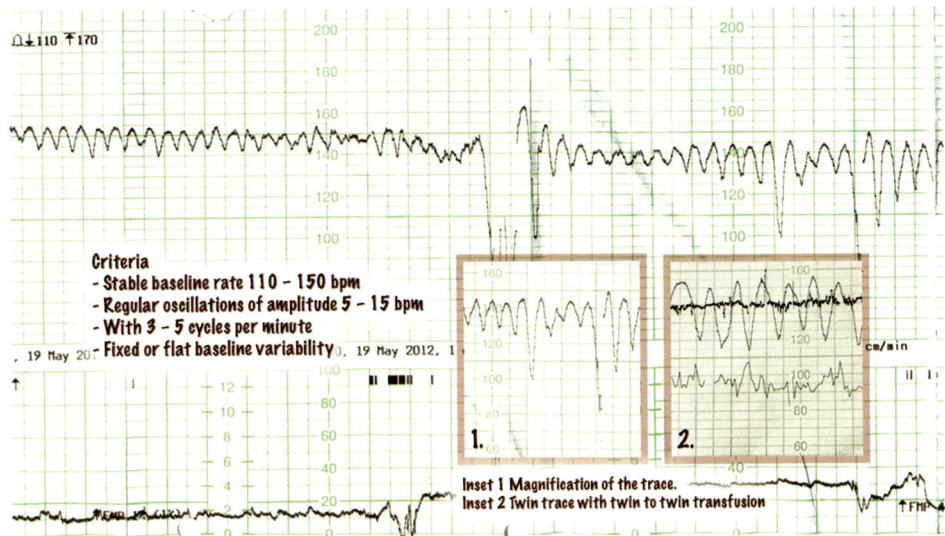

Criteria
- Stable baseline rate 110 – 150 bpm
- Regular oscillations of amplitude 5 – 15 bpm
- With 3 – 5 cycles per minute
- Fixed or flat baseline variability

1. **2.**

Inset 1 Magnification of the trace.
Inset 2 Twin trace with twin to twin transfusion

Fig. 18.7: Sinusoidal pattern.

Sinusoidal pattern is a rare CTG pattern found in fetuses with severe anemia due to any cause. It is a pathological trace criterion and needs rapid identification and action for fetal salvage. Pseudo-sinusoidal or physiological sinusoidal pattern is sometimes seen as a physiological pattern with fetal breathing movements and thumbsucking movements. Actocardiogram, which captures fetal movement in addition to cardiotocography, is helpful to differentiate between them. Pathological sinusoidal trace associated with fetal anemia is shown in Fig. 18.7 to show the differences between the two. The fetal movement perceived in this trace is just 1% (FMP 1%). The inset image 2 is from a twin-to-twin transfusion syndrome pregnancy with pathological trace for both the twins; the anemic fetus throwing a sinusoidal pattern.[8]

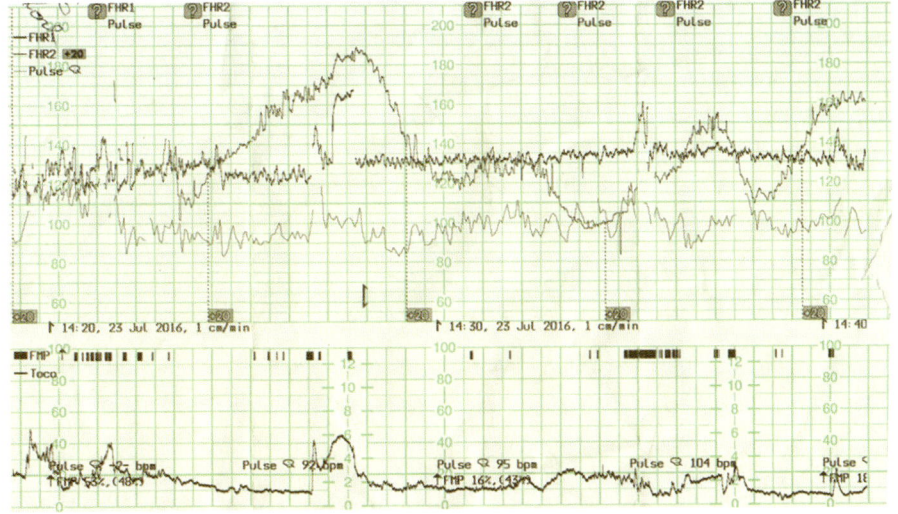

Fig. 18.8: Twin antepartum NST, with discordant fetal growth.

Case 6: Indeterminate Baseline

G2 P1 L1, with 36 weeks gestation, with discordant fetal growth having a NST for antenatal surveillance. First twin had an estimated fetal weight at 23rd centile and the second showed fetal growth restriction at 4th centile. The interpretation of NST of both the twins has to be done with their clinical background (Fig. 18.8). The NST of the twin A is normal. The second twin shows a wandering baseline, difficult to determine, with prolonged decelerations. The decelerations touch 80 bpm, which appear on a printed trace as 100 bpm due to +20 added to the second trace.

CONCLUSIONS

Cardiotocography is a fetal monitoring technique, requires training and experience for better and appropriate use. Obstetric practice should make CTG training and certification mandatory for the staff working in labour wards for CTG has become an extension of clinical examination in high-risk pregnancies more so in critically ill parturients. Regular CTG meetings to review and discuss CTG interpretation and management should be planned to maintain the skills.

REFERENCES

1. National Collaborating Centre for Women's and Children's Health commissioned by the National Institute for Health and Clinical Excellence. Intrapartum care. 2007.

2. Intrapartum care for healthy women and babies (2014 updated 2017) NICE guideline CG190.

3. American College of Obstetricians and Gynecologists. Practice Bulletin No. 145. Antepartum fetal surveillance. Obstet Gynecol 2014;124: 182–92.

4. Aoyama K, Seaward PG, Lapinsky SE. Fetal outcome in the critically ill pregnant woman. Critical Care 2014;18(3):307. doi:10.1186/cc13895.

5. American College of Obstetricians and Gynecologists. ACOG Practice Bulletin No. 116: Management of Intrapartum Fetal Heart Rate Tracings. Obstet Gynecol 2010;116(5):1232–40.

6. Bailey RE. Intrapartum fetal monitoring. Am Fam Physician 2009;80(12):1388–96.

7. Bailey RE. Intrapartum fetal surveillance. In: Leeman L, (Ed). Advanced Life Support in Obstetrics Program: Provider Course Syllabus. Leawood, Kan: American Academy of Family Physicians; 2009.

8. Ayres-de-Campos D, Spong CY, Chandraharan E. The FIGO Intrapartum Fetal Monitoring Expert Consensus Panel. FIGO consensus guidelines on intrapartum fetal monitoring: Cardiotocography. Int J Gynecol Obstet 2015;131:13–24.

The Ominous Sinusoidal Fetal Heart Pattern

Nozer Sheriar, Rajneet Bhatia

Thou ominous and fearful owl of death.

William Shakespeare

There are ominous fetal heart patterns that signify fetal jeopardy. The sinusoidal pattern is one such pattern which is associated with a high-risk fetal morbidity and imminent mortality. It represents an acute disturbance in the regulatory mechanism of Fetal Heart Rate (FHR).

Antenatal and intrapartum fetal monitoring in the form of non stress test and cardiotocography now forms an important part of fetal surveillance. The basic components of FHR assessment–baseline heart rate, baseline variability and heart rate reactivity in response to fetal movements or uterine contractions, signify an intact and functional fetal Central Nervous System (CNS) and denotes fetal well-being. While the Autonomic Nervous System (ANS) directly influences the FHR and its variability. The slowing and acceleration of the FHR along with the short-term and long-term variability is determined by the continuous balance between the parasympathetic and sympathetic nervous systems respectively.[1]

HISTORY AND ANTECEDENTS

Renou et al. recognized the direct influence of the parasympathetic and sympathetic nervous system on the heart rate and its variability. A continuous balance between the parasympathetic and sympathetic nervous system determines the slowing and accelerating FHR, respectively, as well as determining R-R interval differences (short-term variability) and the 2–5 cycles per minute variations of the FHR (long-term variability).[1]

Kubli et al. and Manseau et al. first described an undulating waveform alternating with a flat or smooth baseline FHR in severely affected, Rh-sensitized and dying fetuses.[2,3] Both stated that this particular FHR pattern, whatever its pathogenesis, was an extremely significant finding that implied severe fetal jeopardy and impending fetal death. This FHR pattern was called 'sinusoidal' because of its sine waveform. Subsequently, Modanlou et al. described SHR pattern associated with fetal to maternal hemorrhage causing severe fetal anemia and hydrops fetalis.

DEFINITION AND CHARACTERISTICS

The undulating flow of a heart rate pattern is what defines the sinusoidal heart pattern. However, it is important to appreciate that besides SHR, the undulating pattern may be associated with a number other causes.[4]

1. True SHR pattern
2. Drugs
3. Premortem FHR pattern

4. Pseudo SHR pattern
5. Equivocal FHR patterns.

Modanlou and Freeman proposed the following definitive criteria for the interpretation of true SHR pattern.[5]

1. Stable baseline FHR of 120–160 bpm
2. Amplitude of 5–15 bpm, rarely greater
3. Frequency of 2–5 cycles per minute
4. Fixed or flat short-term variability
5. Oscillation of the sinusoidal wavefrom above and below a baseline
6. No areas of normal FHR variability or reactivity.

PATHOPHYSIOLOGY

The exact cause and pathophysiology of the Sinusoidal Heart Rate (SHR) pattern is not certain but various hypothesis have been proposed. Based on the fact that FHR is a result of the interplay between the sympathetic and parasympathetic nervous systems which are under the control of the higher centers, it has been suggested that sinusoidal patterns occur probably due to the absence of CNS control over the FHR.

The fetal conditions that may have a role in affecting CNS control in causing a sinusoidal pattern are disorders of the autonomic nervous system leading to fetal hypoxia or acidosis or increased cardiac output as a result of chronic intrauterine fetal anemia. There might be associated reactionary myocardial hypertrophy and heart failure depending upon the severity.[6]

Using animal models, an inverse correlation between fetal arterial pH and umbilical cord blood arginine vasopressin level was established. Murata et al. noted a rise of arginine vasopressin levels in post-hemorrhagic or anemic fetal lambs. Further research revealed that with chemical or surgical vagotomy, arginine vasopressin infusion produced SHR pattern, thus suggesting the role of ANS dysfunction combined with the increase in arginine vasopressin as the causative condition.[7]

This rise in arginine vasopressin level drops the fetal pH and is directly or indirectly responsible for SHR pattern in anemic and/ or acidotic fetuses.[8] Additionally, it was proposed that absent or decreased vagal efferent control of the fetal heart is needed for SHR pattern induced by arginine vasopressin. It is also suggested that alternating hypervolemia and hypovolemia caused by umbilical cord compression could result in SHR pattern.

SHR patterns have been detected in preterm infants with periodic breathing and during neonatal generalized seizure activity. Periodic breathings are known to be associated with CNS hypoxia which may also occur with generalized seizure activities. These observations also suggest that true SHR pattern is associated with hypoxemia and tissue hypoxia of the CNS and ANS dysfunction.

SINUSOIDAL VS PSEUDOSINUSOIDAL PATTERNS

The true sinusoidal pattern is rare but ominous. It is a regular, smooth, undulating form typical of a sine wave that occurs with a frequency of two to five cycles per minute and an amplitude range of 5 to 15 bpm. It is also characterized by a stable baseline heart rate of 120 to 160 bpm and absent beat-to-beat variability. It should be differentiated from the pseudosinusoidal pattern (Fig. 19.1A), which is a benign, uniform long-term variability pattern. A pseudosinusoidal pattern shows less regularity in the shape and amplitude of the variability waves and the presence of beat-to-beat variability, compared with the true sinusoidal pattern[9] (Fig. 19.1B).

Although detection of fetal compromise is the purpose of fetal monitoring, there are the associated risks of false-positive tests that may lead to unnecessary surgical intervention. Since sinusoidal patterns require emergency intrauterine fetal resuscitation and immediate delivery it is important to differentiate these from pseudosinusoidal patterns to guide appropriate triage decisions.

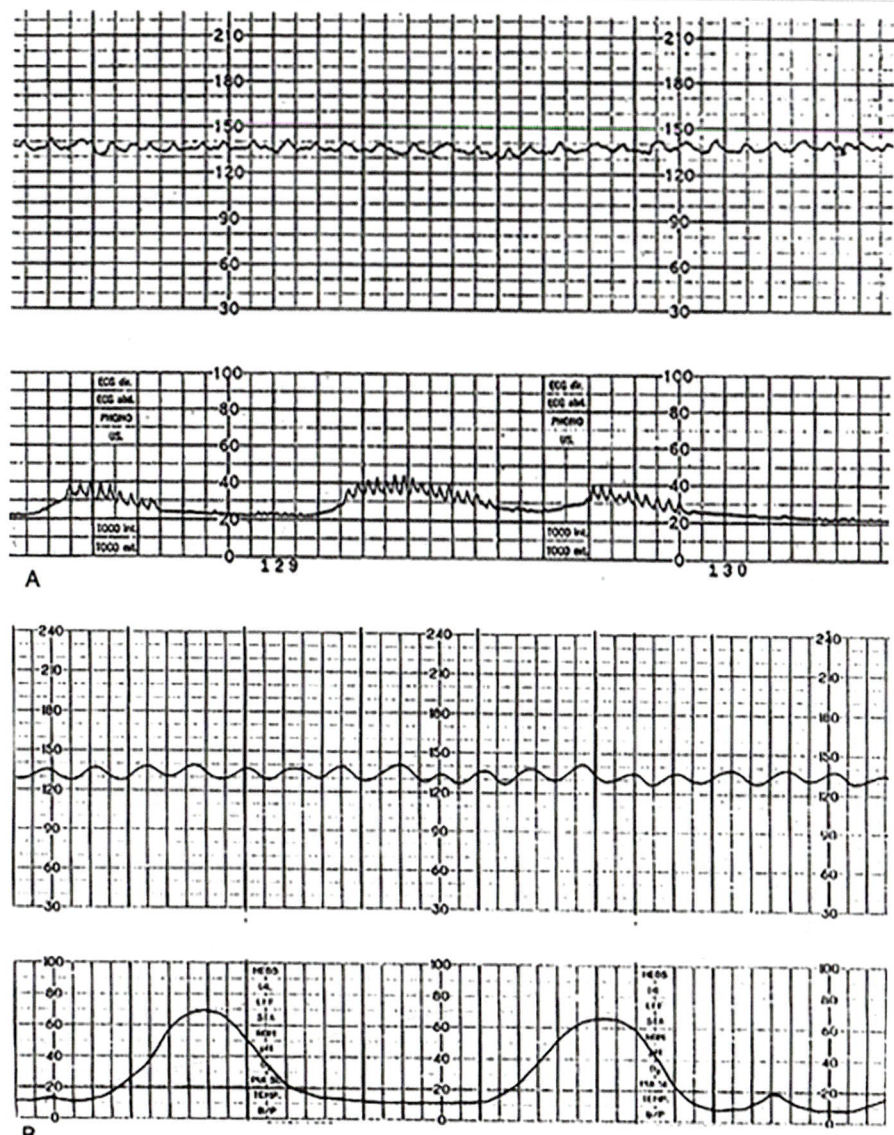

Figs 19.1A and B: Diagrammatic representation comparing sinusoidal with pseudosinusoidal patterns.[8] (A) Pseudosinusoidal pattern with decreased regularity and the preserved beat-to-beat variability. (B) Sinusoidal pattern with persistent regular undulations and loss of beat-to-beat variability.

Hofmeyr and Sonnendecker critically reviewed reports of sinusoidal patterns in literature and proposed stricter criteria by which the sinusoidal pattern can be diagnosed and distinguished from the pseudosinusoidal pattern.[10]

CLINICAL SIGNIFICANCE

Patterns characterized by a loss in baseline variability followed either by an instability of the heart rate as manifested by bradycardia or the occurrence of sinusoidal pattern had the highest association with intrauterine fetal deaths. This is why true sinusoidal patterns are labelled to be ominous and have poor fetal and neonatal outcome unless emergency measures are taken.

In contrast the pseudosinusoidal pattern is also seen in certain cases in which the sine waves show less regularity in shape and amplitude with the presence of normal baseline variability. This is the benign pattern in which the fetal and neonatal outcome is not compromised. Hence, to avoid unnecessary invasive intervention one reading alone of sinusoidal pattern should not be a reason for emergency obstetric measure. Instead the patient should be under surveillance and investigated further and the possibility of maternal intake of drugs (narcotics and sedatives) for pain relief during labour must be ruled out.

Fetal biophysical profile and MCA Doppler forms the next step in management. Any abnormality in the biophysical profile may point towards fetal jeopardy. Increased pulsatility index in the MCA Doppler may be an indicator of fetal anemia. Antenatal Kleuiher Betke's test is also suggestive of feto-maternal hemorrhage if positive. Moreover, a persistent sinusoidal pattern in the absence of any positive clinical or ultrasound finding can also be considered for emergency obstetric management in a setting of good neonatal intensive care facilities.[11]

CAUSES AND ASSOCIATED CONDITIONS

A number of fetal conditions have been reported to cause SHR. Some of these are ominous and others innocuous.[12]

Physiological Causes

- Rhythmic movements of the fetal mouth and suckling
- Fetal breathing movements (may be related to fetal respiratory arrhythmia)
- Fetal sleep cycles (more in NREM sleep than REM sleep)

- Effect of narcotic drugs; alphaprodine, meperidine and butorphanol.

Pathological Causes

- Severe fetal anemia due to Rh isoimmunization
- Massive fetomaternal hemorrhage
- Severe fetal hypoxia and asphyxia
- Fetal intracranial hemorrhage
- Twin to twin transfusion syndrome
- Traumatic fetal bleeding
- Vasa previa with bleeding.

Besides these sinusoidal patterns have been reported with congenital hydrocephalus and gastroschisis, after traumatic amniocentesis and during maternal cardiopulmonary bypass.[13–15]

CLINICAL PRESENTATION THROUGH ILLUSTRATIVE CASES

Case 1: Primigravida at 39 weeks gestation presented with PROM and a history of loss of fetal movements since 3 days. Her antenatal period was uneventful. On examination she was 3 cm dilated with absent membranes and meconium stained liquor. The admission test was performed (Fig. 19.2) which was suggestive of sine wave like undulating pattern with variability of under 4 beats per minute with a cycle frequency of 3 per minute. There were variable decelerations to 80 beats per minute.

In view of the sinusoidal pattern with decelerations and meconium stained liquor, emergency cesarean was performed and 2.35 kg baby with severe pallor was delivered with heart rate of 124 beats and a respiratory rate of 32 breaths per minute. The neonate was shifted to NICU, the hemoglobin being 6 gm% with a Kliehauer test revealing a 3% feto-maternal leak, corresponding to approximately 150 ml. Intensive care management was done along with transfusion of 122 ml of packed cells. Baby was discharged on day 14 with mother.[16]

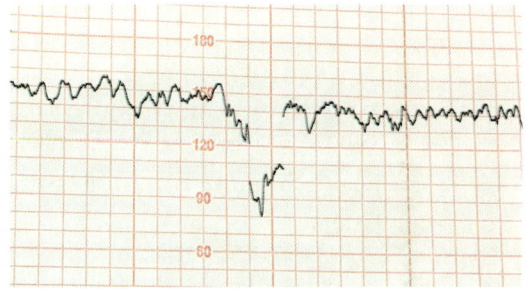

Fig. 19.2

Case 2: Primigravida with 36 weeks gestation presented with decreased fetal movements since 6 hours. Her antenatal period was uneventful. On examination she was vitally stable with abdominal findings corresponding to the weeks of gestation. Per vaginally os was closed with poorly effaced cervix. Nonstress test was performed (Fig. 19.3) which was suggestive of sine wave like undulating pattern of amplitude of 5–15 beats with a cycle frequency of 3–4 per minute. Biophysical profile was done which was normal.

MCA Doppler was suggestive of normal pulsatility index for that gestation. The NST was repeated after 24 hours which was suggestive of persistent sinusoidal pattern (Fig. 19.4).

In view of the persistent sinusoidal pattern, emergency cesarean was performed and 2.1 kg baby with severe pallor was delivered with Apgar score of 7 and 8 at 1 and 5 minutes respectively. The neonate was shifted to NICU, the hemoglobin being 4 gm% with hematocrit of 12% and went in shock after 4 hours, which required intubation and

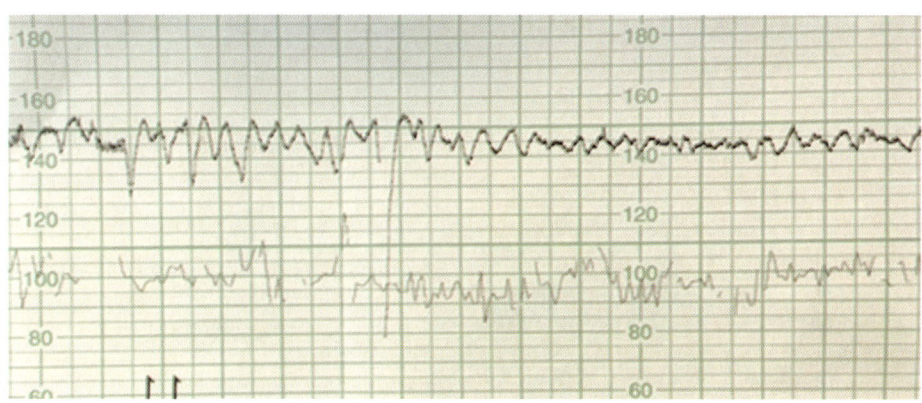

Fig. 19.3

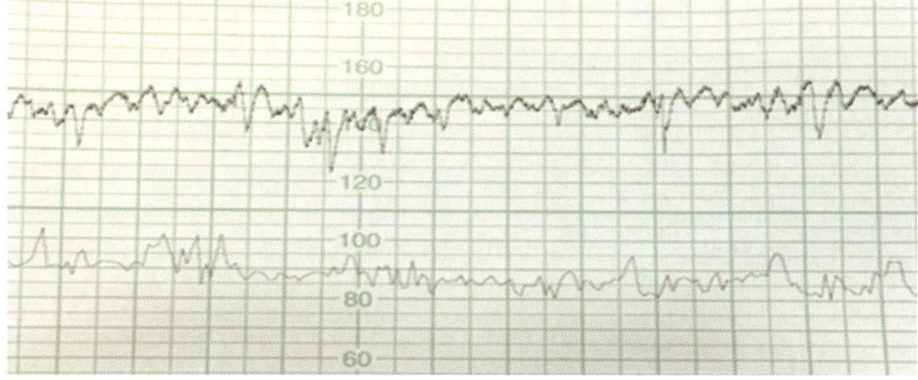

Fig. 19.4

ventilation. Intensive care management was done along with transfusion of packed cells. Postoperative calculation using changes in HbF levels and correlation with estimated drop in fetal hemoglobin, estimated approximately 290 ml of fetomaternal hemorrhage. Baby was discharged on day 8 with mother.[17]

CONCLUSION

True SHR pattern is a rare occurrence but ominous sign and signifies severe fetal anemia and hypoxia or asphyxia. Intake of narcotics and sedatives should be ruled out before labelling the pattern as sinusoidal. A normal pattern before or after the SHR pattern rules out the diagnosis of true SHR pattern. We suggest that only one reading of sinusoidal pattern should not be an indication for intervention but decision must be taken by considering clinical parameters, persistent sinusoidal pattern and fetal biophysical profile with MCA doppler. That having been said, identification of this ominous pattern and timely intervention is a life-saving opportunity.

REFERENCES

1. Renou P, Warwick N, Wood C. Autonomic control of fetal heart rate. Am J Obstet Gynecol. 1969;105:949.

2. Kubli F, Ruttgers H, Haller U, et al. The antepartum fetal heart rate; II. Rate, variability and deceleration. Z Gerburtshilfe Perinatol. 1972; 176:309.

3. Manseau P, Vaquier J, Chavinie J and Sureau C. Le rythmeardiaque foetal sinusoidal. Aspect evocateur de souffrancefoetale au cours de la grossesse. J Gynecol Obstet Biol Reprod1972; 1:343.

4. Modanlou HD, Murata Y. Sinusoidal heart rate pattern: Reappraisal of its definition and clinical significance. J Obstet Gynaecol Res 2004; 30(3):169.

5. Modanlou HD, Freeman RK. Sinusoidal fetal heart rate pattern. Its denition and clinical signicance. Am J Obstet Gynecol 1982;142:1033.

6. Graca LM, Cardoso CG, Calhaz-Jorge C. An approach to interpretation and classification of sinusoidal fetal heart rate pattern. Eur J Obstet Gynecol Reprod Biol 1988;21:203.

7. Murata Y, Miyake Y, Yamamato T, et al. Experimentally produced sinusoidal fetal heart rate pattern in the chronically instrumented fetal lamb, Am J Obstet Gynecol 1985;153:693.

8. Treisser P, Vige B, Maria F, et al. Sinusoidal fetal heart rate pattern in severe fetal anemia from fetomaternal transfusion. Int J Gynaecol Obstet, 1982; 20:211.

9. Sweha A, Hacker TW, Nuovo J. Interpretation of the electronic fetal heart rate during labor, Am Fam Physician 1999;59(9):2487.

10. Hofmeyr GJ, Sonnendecker EW. Pseudo-sinusoidal fetal heart rate patterns, S African Med J 1983;64(1):19.

11. Yambao TJ, David Clark D, Ashby W, Raja Abdul-Karim R. Sinusoidal fetal heart rate patterns. Case reports and management. Br J Obstet Gynaecol, 1982;89:765.

12. Schneider EP, Tropper PJ. The variable deceleration, prolonged deceleration, and sinusoidal fetal heart rate. Clin Obstet Gynecol 1986;29:64.

13. Ombelet W, Van der Merwe JV. Sinusoidal fetal heart rate pattern associated with congenital hydrocephalus. A report of 2 cases. S African Med J 1985;67:423.

14. Elliott JP, Castro RJ, O'Keefe DF. Sinusoidal fetal heart rate pattern associated with gastroschsis. Am J Perinatol 1988;5:295.

15. Burke AB, Hur D, Bolan JC, et al. Sinusoidal fetal heart rate patter during cardiopulmonary bypass. Am J Obstet Gynecol 1990;163:17.

16. Sheriar NK, Dastur AE, Shah D, Purandare CB. Chronic fetal anemia due to spontaneous fetomaternal hemorrhage. J Obstet Gynecol India, 1989;39:735.

17. Sheriar NK, Bhatia RN. The rare sinusoidal pattern: A life saving trace. 45th Annual MOGS conference, 2017.

Importance of CTG in Cerebral Palsy

Achla Batra, Monika Gupta

Delivering a healthy fetus in all aspects is the ultimate goal of the labouring mother and the whole obstetrical science at her help. Medical science has achieved a lot in reducing the unpredictability which can put both maternal and fetal health at danger. CTG and intermittent auscultation are the two methods most reliably used for fetal monitoring in labour, with more emphasis on CTG these days due to the convenience it provides.[1]

There has been a debate involving obstetricians, pediatricians, parents and lawyers regarding the relation between the quality of care given to a mother during labour and delivery and the death of her baby or cerebral palsy in her surviving child. Currently, CTG has been put under scanner due to its low sensitivity to detect fetal distress and low positive predictive value for future neurological disorder in fetus. Its high false positive rates have increased the rate of operative deliveries.

CEREBRAL PALSY

Cerebral Palsy (CP) refers to a heterogeneous group of conditions involving permanent non-progressive central motor dysfunction that affect muscle tone, posture, and movement. The birth prevalence of cerebral palsy shows geographical variations, but is generally in the range of 1.5–3 per 1000 live births.[2]

The prevalence of CP is far higher in preterm compared with term infants, and increases with decreasing gestational age and birth weight as this conditions is due to abnormalities of the developing fetal or infantile brain resulting from a variety of causes.

Cerebral palsy was once thought to be caused mainly by birth asphyxia. If the infant developed CP, cognitive deficits, epilepsy, or any other deficits or impairment, then the cause was assigned as asphyxia and cure was quick delivery. Thus, intermittent auscultation became the standard of care. And assisted deliveries became standard for any suspected fetal asphyxia.[3]

CTG came in practice in 1960 and early advocates of electronic fetal monitoring expected that early detection of asphyxia in labour by CTG would largely eliminate cerebral palsy. Studies using surrogate end points such as low neonatal pH and low Apgar scores encouraged that expectation.[4]

During the past decade it has become clear that, contrary to previous views, birth asphyxia does not account for the overwhelming number of cases of cerebral palsy.[5] The actual percentage reported in the literature is quite low. In fact other, interauterine events and the sequelae of preterm birth make the largest contributions to the overall prevalence.

Epidemiological analysis has defined a large number of risk factors for cerebral palsy, including low birth weight, preterm birth, multiple pregnancy, neurological disorder in mother or sibling, thyroid disease or therapy during pregnancy, thyroid hormone deficiency in preterm infants, low placental weight, chorioamnionitis, birth asphyxia, and neonatal hyperbilirubinemia.[6]

Criteria which Associate CP with Perinatal Birth Asphyxia

Report of the American College of Obstetricians and Gynecologists' Task Force on Neonatal Encephalopathy (released in 2003,

updated in 2014) set out certain criteria (Table 20.1) to associate peripartum asphyxia as a cause for any neurological damage.[7]

In a study using the criteria set forth by the American College of Obstetricians and Gynecologists (ACOG) and the International Cerebral Palsy Task Force, an acute intrapartum hypoxic event was identified in only 1% (2 of 213 infants) of children with CP.

Can cerebral palsy be prevented by use of electronic fetal monitoring?
It has been surprisingly difficult to show that complex modern fetal monitoring reduces the

---

Table 20.1: Markers of an acute peripartum or intrapartum hypoxic-ischemic event

Neonatal signs consistent with an acute peripartum or intrapartum event

- Apgar score of <5 at 5 minutes and 10 minutes
- Fetal umbilical artery acidemia: fetal umbilical artery pH <7.0, or base deficit ≥ 12 mmol/L, or both
- Neuroimaging evidence of acute brain injury seen on brain MRI or MRS consistent with hypoxia-ischemia
- Presence of multisystem organ failure consistent with hypoxic-ischemic encephalopathy

Type and timing of contributing factors that are consistent with an acute peripartum or intrapartum events

Type and timing of contributing factors that are consistent with an acute peripartum or intrapartum events:
1. A sentinel hypoxic or ischemic event occurring immediately before or during labour and delivery:
   - Ruptured uterus
   - Severe abruptio placentae
   - Umbilical cord prolapse
   - Amniotic fluid embolus with coincident severe and prolonged maternal hypotension and hypoxemia
   - Maternal cardiovascular collapse
   - Fetal exsanguination from either vasa previa or massive fetomaternal hemorrhage
2. Fetal heart rate monitor patterns consistent with an acute peripartum or intrapartum event, particularly a category I fetal heart rate pattern on presentation that converts to one of the following patterns:
   - Category III pattern
   - Tachycardia with recurrent decelerations
   - Persistent minimal variability with recurrent decelerations
3. Timing and type of brain injury patterns based on imaging studies consistent with an etiology of an acute peripartum or intrapartum event. Well-defined patterns on brain MRI typical of hypoxic-ischemic cerebral injury in the newborn are:
   - Deep nuclear gray matter (i.e. basal ganglia or thalamus) injury
   - Watershed (borderzone) cortical injury
   - No evidence of other proximal or distal factors that could be contributing

Developmental outcome is spastic quadriplegia or dyskinetic cerebral palsy

- Other subtypes of cerebral palsy are less likely to be associated with acute intrapartum hypoxic-ischemic events
- Other developmental abnormalities may occur, but they are not specific to acute intrapartum hypoxic-ischemic encephalopathy and may arise from a variety of other causes

incidence of fetal death or neonatal encephalopathy, and the value of fetal monitoring for predicting events leading to cerebral palsy is uncertain. Fetal heart rate decelerations are commonly associated with peripheral chemoreflex rather than compression of the fetal head or umbilical cord, and may be relatively unthreatening.[8] Adding additional observations, including depression of the ST segment of the fetal electrocardiogram, has not improved reliability or improved neonatal outcome.[9]

Another fact, questioning the efficacy of CTG in preventing CP, is that the prevalence of CP has not decreased in developed countries over the past 30 years, despite the widespread use of electronic FHR monitoring and a 5- to 6-fold increase in the cesarean delivery rate.[10] Also notable is the epidemiological observation that Hypoxic-Ischemic Encephalopathy (HIE) due to birth asphyxia at term accounts for between 10 and 30% of cerebral palsy in developed countries[6] which is similar to estimates in the developing world, where CTG monitoring is less commonly available.[11]

Electronic fetal monitoring has also not been able to lower the rates of perinatal death, intrapartum stillbirth, neonatal death, low or very low Apgar scores, need for special neonatal care, or neonatal death.[12]

A statistical reduction in the incidence of cerebral palsy with continuous intrapartum FHR monitoring is also difficult to prove for several reasons. Many cases of cerebral palsy are due to antepartum, rather than intrapartum, events. In such cases, intrapartum interventions are unlikely to change the course of the disease. Most FHR abnormalities are not associated with fetal acidemia or hypoxemia, and most fetal acidemia and hypoxemia does not result in neurologic disability.

Abnormal FHR patterns have a high false-positive rate and are poor predictors of the subsequent development of cerebral palsy. For intrapartum asphyxia to develop in a previously normal fetus, an acute sentinel event must occur. Most fetuses with abnormal fetal heart tracings that present to the hospital may have already experienced damage, and prompt delivery by cesarean section may not improve the outcome.[7]

According to Cochrane meta-analysis reported in 2017, CTG during labour is associated with reduced rates of neonatal seizures, but no clear differences in cerebral palsy, infant mortality or other standard measures of neonatal well-being. However, continuous CTG was associated with an increase in cesarean sections and instrumental vaginal births. In fact, one study calculated that 99.8% of "abnormal" FHR tracings are not associated with the later development of cerebral palsy, yielding a positive predictive value that is equal to, or lower than, the prevalence of cerebral palsy in the general population.[1]

A meta-analysis evaluating intrapartum electronic FHR monitoring and the prevention of perinatal brain injury which included studies that quantified intrapartum EFM and its relation to specific neurologic outcomes (seizures, periventricular leukomalacia, cerebral palsy, death) concluded that although intrapartum EFM abnormalities correlate with umbilical cord base excess and its use is associated with decreased neonatal seizures, it has no effect on perinatal mortality or pediatric neurologic morbidity.[13] Inter-intraobserver interpretation variability of EFM is exceedingly high and reinterpretation by the same observer at a later occasion has long been known to be highly biased and unreliable.[14]

The recent debate is between the use of intermittent auscultation and EFM for intrapartum fetal monitoring to predict birth asphyxia. There are no randomized trials of intrapartum fetal monitoring versus no intrapartum fetal monitoring, as such kind of trials would be not be ethically appropriate.

Only two randomised trials, both published more than 25 years ago, have compared cerebral palsy rates in births monitored electronically or by intermittent auscultation. Cerebral palsy cannot be diagnosed at birth, so infants have to be followed for several years, until a diagnosis can reliably be made. In a trial in 13079 births in Dublin, evaluation at age 4 years indicated that the rate of cerebral palsy was not lower in children whose births were monitored electronically.[15]

Another multicentre randomised trial in the US compared 93 singleton, vertex presenting infants with birth weight ≤ 1750 g whose births were monitored electronically with 96 comparable children monitored by intermittent auscultation. At 18 months of age, the cerebral palsy rate was significantly higher in the electronically monitored group.[16]

Evidence-Based Recommendations for EFM

There are still lesser evidence to put the CTG out of practice as the intermittent auscultation is not practical in every situation, moreover, few of the advantages given by the use of CTG would be overlooked. Also, practising in the current time period when the medicolegal litigations are its high, CTG provides a substantial proof for the status of the *in utero* fetus. It is one of the important prescribed factor amongst the criteria set for cerebral palsy; though it may be able to assess the risk of cerebral palsy in only a small fraction of cases amongst those having intrapartum birth asphyxia.

Instead of using any particular method in isolation a protocol should be designed to provide wholesome care to the fetus and detect any abnormality, as advised by many global organisations:

- **American College of Obstetricians and Gynecologists**, 2009:[17]
  - Either continuous electronic FHR monitoring or intermittent auscultation is acceptable in uncomplicated patients.
  - High-risk pregnancies (e.g. pre-eclampsia, suspected growth restriction, type I diabetes mellitus) should be monitored continuously during labour.

- **National Institute for Health and Care Excellence, 2014:**[18]
  - In all birth settings, offer intermittent auscultation to low-risk women in the first stage of labour. Do not perform cardiotocography in low-risk women.
  - Advise continuous cardiotocography if any of the following risk factors occur during labor:
    - Suspected chorioamnionitis, sepsis, or temperature $\geq 38°C$
    - Severe hypertension ($\geq 160/110$ mm Hg)
    - Oxytocin use
    - Significant meconium
    - Fresh vaginal bleeding.
- If continuous cardiotocography was used because of concerns arising from intermittent auscultation but the tracing is normal after 20 minutes of observation, remove the cardiotocograph and return to intermittent auscultation.

Also while using CTG following principles should be kept in mind:

1. Confirm that the monitor is recording the FHR and uterine activity adequately to permit appropriately-informed management decisions. Ensure that the maternal heart rate is not being recorded.
2. Assess uterine activity along with baseline FHR, variability, accelerations, decelerations, and sinusoidal pattern, and place the tracing in a category I, II, or III.
3. If the tracing is category I and the patient is low-risk, initiate routine intrapartum fetal surveillance.
4. If the tracing is not category I, evaluate the integrity of the fetal oxygen pathway (maternal lungs, heart, and vasculature, as well as the uterus, placenta, and umbilical cord). Attempt to correct the problem, if possible, by initiating measures to improve fetal oxygenation, such as reduction in dose or discontinuation of oxytocin, maternal repositioning, and intravenous fluid bolus. Also employ ancillary tests to further assess

the fetal condition (fetal scalp stimulation, fetal ST analysis, fetal cord blood analysis, fetal pulse oximetry).

If the FHR pattern does not improve within a reasonable period of time, begin planning for the possible need for rapid delivery. This may include availability of an operating room and specialized equipment, notification of anesthesia and pediatrics, consent forms, and laboratory tests.

CTG, CP and Litigation

The cardinal driver of cerebral palsy litigation is electronic fetal monitoring, which has continued unabated for 40 years because of its high false positive rate.[19] The question of proper use of electronic fetal monitoring arises in most cerebral palsy claims despite the evidence that it has high false positive rate and such monitoring has not been successful in prevention of cerebral palsy.[4]

It should be made known to all medical professionals, lawyers and public that except in rare instances, cerebral palsy is a developmental event that is unpreventable given our current state of technology. The criteria set forth by the American College of Obstetricians and Gynecologists (ACOG) and the International Cerebral Palsy Task Force[7] should be taken into account before fixing accountability of CP to intrapartum asphyxia.

CONCLUSION

CTG has low sensitivity to detect the fetal distress and low positive predictive value for future neurological disorder or cerebral palsy in fetus. There is some association between suboptimal care and cerebral palsy, but this seems to have a role in only a small proportion of all cases of cerebral palsy. The contribution of adverse antenatal factors in the origin of cerebral palsy needs further study.

An integrated protocol using the bests of all the available methods should be used and assessment of the maternal and fetal condition in total should not be overlooked.

FUTURE RESEARCH

A new metric: the "Fetal Reserve Index" (FRI), formally incorporating EFM with maternal, obstetrical, fetal risk factors, and excessive uterine activity for assessment of risk for CP has been presented. Eden et al. in a retrospective trial reported that an abnormal FRI identified all cases of labour-related neurological injury more reliably and earlier than Category III CTG, which may allow fetal therapy by intrauterine resuscitation.[20] This quantified screening system needs further evaluation in prospective trials.

REFERENCES

1. Alfirevic Z, Devane D, Gyte GM, Cuthbert A. Continuous cardiotocography (CTG) as a form of electronic fetal monitoring (EFM) for fetal assessment during labour. Cochrane Database Syst Rev. 2017;2:CD006066.
2. Paneth N, Kiely J. The frequency of cerebral palsy: a review of population studies in industrialized nations since 1950. In: Stanley F, Alberman E (Eds). The epidemiology of the cerebral palsies. Blackwell Scientific, Oxford 46–56.
3. Sartwelle TP. Electronic fetal monitoring: a bridge too far. J Legal Med 2012;33:313–79.
4. Nelson KB, Sartwelle TP, Rouse DJ. Electronic fetal monitoring, cerebral palsy, and caesarean section: assumptions versus evidence. BMJ 2016;355:i6405.
5. Adamson SJ, Alessamdri AM, Badawi N, Burton PR, Pemberton PJ, Stanley F. Predictors of neonatal encephalopathy in full term infants. BMJ 1995;311:598–602.
6. Lee AC, Kozuki N, Blencowe H, et al. Intrapartum-related neonatal encephalopathy incidence and impairment at regional and global levels for 2010 with trends from 1990. Pediatric Research 2013;74(Suppl 1):50–72. doi:10.1038/pr.2013.206.
7. Executive summary: Neonatal encephalopathy and neurologic outcome, second edition. Report of the American College of Obstetricians and Gynecologists' Task Force on Neonatal Encephalopathy. Obstet Gynecol 2014;123(4):896–901. doi: 10.1097/01.AOG.0000445580.65983.d2.
8. Lear CA, Galinsky R, Wassink G, et al. The myths and physiology surrounding intrapartum decelerations: the critical role of the peripheral

chemoreflex. J Physiol 2016;594:4711–25. doi:10.1113/JP271205 pmid:27328617.

9. Bloom SL, Belfort M, Saade G. Eunice Kennedy Shriver National Institute of Child Health and Human Development Maternal-Fetal Medicine Units Network. What we have learned about intrapartum fetal monitoring trials in the MFMU Network. Semin Perinatol 2016;40:307–17. doi:10.1053/j.semperi.2016.03.008 pmid:27140936

10. Boog G. Cerebral palsy and perinatal asphyxia (I—diagnosis). Gynecol Obstet Fertil. 2010;38(4):261–77. doi:10.1016/j.gyobfe.2010.02.009.

11. Ellis M, Manandhar N, Manandhar DS, Costello AM de L. Risk factors for neonatal encephalopathy in Kathmandu, Nepal, a developing country: unmatched case-control study. BMJ 2000;320(7244):1229–36.

12. Alfirevic Z, Devane D, Gyte GM. Continuous cardiotocography (CTG) as a form of electronic fetal monitoring (EFM) for fetal assessment during labour. Cochrane Database Syst Rev 2013;5:CD006066.pmid:23728657.

13. Graham EM, Petersen SM, Christo DK, Fox HE. Intrapartum electronic fetal heart rate monitoring and the prevention of perinatal brain injury. Obstet Gynecol 2006;108(3 Pt 1):656–66.

14. Blackwell SC, Grobman WA, Antoniewicz L, Hutchinson M, Gyamfi Bannerman C. Inter-observer and intraobserver reliability of the NICHD 3-Tier Fetal Heart Rate Interpretation System. Am J Obstet Gynecol 2011;205(4):378.e1–5.

15. Grant A, O'Brien N, Joy MT, Hennessy E, MacDonald D. Cerebral palsy among children born during the Dublin randomised trial of intra-partum monitoring. Lancet 1989;2:12336. doi:10.1016/S0140-6736(89)91848–5.

16. Shy KK, Luthy DA, Bennett FC, et al. Effects of electronic fetal-heart-rate monitoring, as com-pared with periodic auscultation, on the neuro-logic development of premature infants. N Engl J Med 1990;322:588–93.

17. American College of Obstetricians and Gyne-cologists. ACOG Practice Bulletin No. 106: Intra-partum fetal heart rate monitoring: nomenclature, interpretation, and general management principles. Obstet Gynecol 2009;114:192–202.

18. NICE. Intrapartum care of healthy women and their babies during childbirth. Draft for consulta-tion, March 2014. www.nice.org.uk/nicemedia/live/13511/67645/67645.pdf.

19. Cerebral Palsy Litigation: Change Course or Abandon Ship Thomas P. Sartwelle, BBA, LLB1 and James C. Johnston, MD, JD2 Journal of Child Neurology 2015;30(7):828–41.

20. Eden RD, Evans MI, Evans SM, Schifrin BS. The Fetal Reserve Index: Re-Engineering the Inter-pretation and Responses to Fetal Heart Rate Patterns. Fetal Diagn Ther 2017;doi: 10.1159/000475927.

Importance of CTG in Fetal Acidosis

Pratima Mittal, Divya Pandey

Electronic Fetal Monitoring (EFM) is a surveillance tool to detect hypoxia or acidosis which can lead to fetal brain injury in a normal or an already compromised fetus.

During labour, the fetuses are on a *continuum of casualty*. In event of persistent decreased oxygenation, a well or non-hypoxic fetus can gradually become hypoxemic, hypoxic or acidotic and ultimately asphyxiated (Fig. 21.1).[1] Fetal asphyxia is thus a sequel of hypoxemia progressing to hypoxia or acidosis. Fetus in early stages of hypoxemia responds well due to intact defense system (Fig. 21.2). When the low oxygen condition persists, the response is blunted and there follows a state of fetal hypoxia which is a stage of limited compensation which the fetus can tolerate only for few hours (Fig. 21.3). However, with the ongoing decreased oxygen levels, there ensues a stage of asphyxia and decompensation which warrants immediate delivery to salvage the fetus (Fig. 21.4).

Fetal Hypoxia is thus defined as a condition where there is a decrease in oxygen concentration in fetal tissues which is inadequate to maintain normal cell energy production by

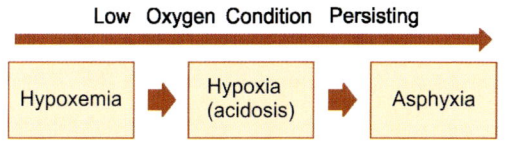

Fig. 21.1: Fetal response to persisting low level of oxygen.

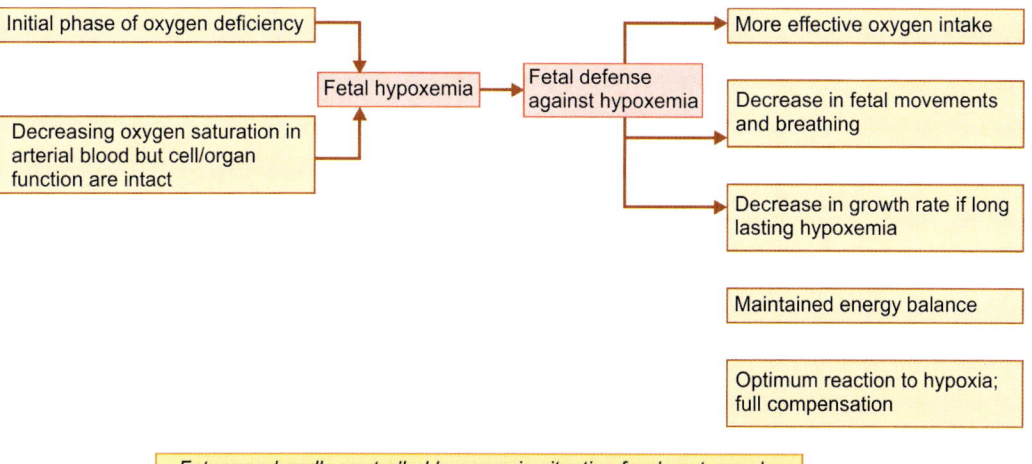

Fig. 21.2: Fetal response to hypoxemia.

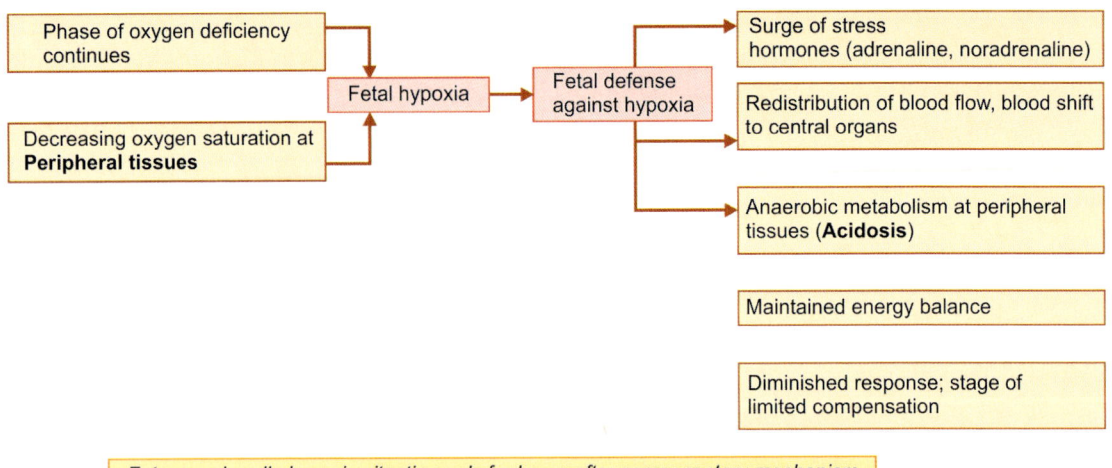

Fig. 21.3: Physiology of fetal hypoxia (acidosis).

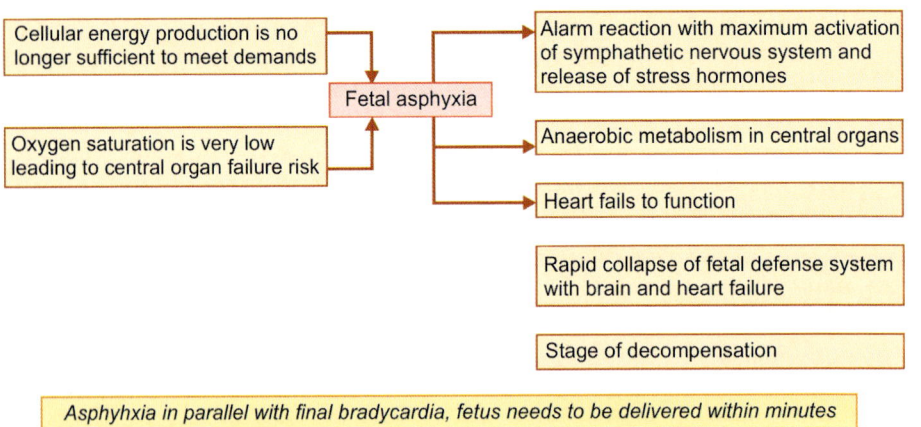

Fig. 21.4: Fetal response to fetal Asphyxia.

aerobic metabolism. *Acute fetal hypoxia* is a condition with acute and fast reduction in oxygen levels occurring over a course of few minutes. As a result of fetal hypoxia, the fetus resorts to anaerobic metabolism which yields energy (less than in aerobic metabolism) and results in lactic acid production. Thus, there is rise of hydrogen ions at intra and extracellular level causing *metabolic acidosis. Fetal hypoxia and acidosis go hand in hand.* These H⁺ ions when released in circulation lead to *metabolic acidemia.* Metabolic acidosis is thus defined as a pH less than 7.00, base deficit in excess of 10 mmol/l or lactate value in excess of 10 mmol/l.[2]

Compensatory/Defense Mechanisms in Fetus

For neutralizing hydrogen ions, there are bases (bicarbonates, hemoglobin and plasma proteins) in fetal blood. As they have limited availability, their depletion (base deficit) is directly related to metabolic acidosis severity. Increasing H⁺ concentration due to base depletion affects cell homeostasis eventually leading to cell death. When hypoxia is extreme

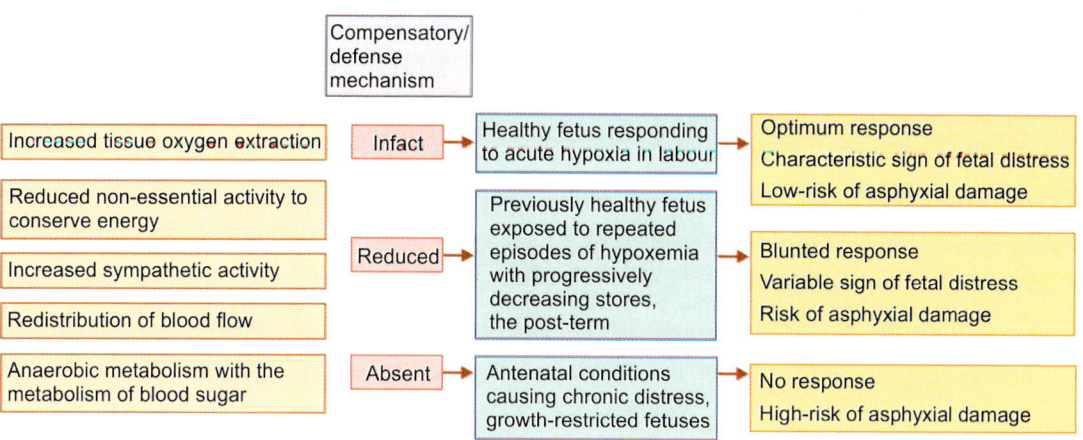

Fig. 21.5: Fetal compensatory/defense mechanisms to handle oxygen deficiency.

and protracted, neurological, respiratory and cardiovascular control can be reflected in decreased APGAR scores at birth (Fig. 21.5).

Magnitude of Problem of Fetal Hypoxia

The overall incidence of fetal hypoxia/acidosis has been defined by incidence of newborn metabolic acidosis varies from 0.06 to 2.8%.[3]

Main Causes, Risk Factors and Consequences of Fetal Hypoxia

The reversible causes include uterine hyper-contractility, sudden maternal hypotension due to supine position of the mother causing aortocaval compression, whereas the irreversible causes are, abruption, uterine rupture and cord prolapse. Maternal cardio-respiratory disorders like severe asthma, shock, pulmonary embolism, cardio-pulmonary arrest and generalized convul-sions may also lead to fetal hypoxia. There are some occult causes too such as cord compression as in true knot, nuchal cord with stretching and major fetal hemorrhage like rupture vasa previa. Some mechanical intra-partum complications can also be accounted for this condition like, shoulder dystocia and arrest of after coming head of the breech. Underlying risk factors for these conditions are induction or augmentation of labour using prostaglandins or oxytocin leading to uterine hypercontractility, regional analgesia causing sudden maternal hypotension and early amniotomy leading to hypercontractility and cord prolapse.[3]

Although most newborns with metabolic acidosis and low Apgar scores, recover, but in some where the hypoxia is extreme and prolonged, neurological dysfunction may ensue over first 48 hours of life manifested by hypotonia, convulsion and/or coma, a situa-tion called Hypoxic-Ischemic–Encephalo-pathy (HIE). Cerebral palsy of dyskinetic or spastic quadriplegia is the long-term neuro-logical sequel most strongly associated with fetal hypoxia, although only 10–20% of these cases are caused by hypoxia.[3]

Thus, it is very important to predict and diagnose the fetuses at risk so as to avert these potentially threatening short-term and long-term sequelae.

Cardiotocographic Changes in Response to Hypoxia

EFM or cardiotocography can be helpful in diagnosing changes and fetal response to hypoxia. Broadly, in response to acute hypoxia, chemoreceptors in fetus get activated and stimulate both sympathetic and para-sympathetic activity. As a result, the **initial response to acute hypoxia** is a drop in fetal heart rate, i.e. **bradycardia**. However, gradually developing or uniformly maintained fetal hypoxia leads to an increase in fetal heart rate.

1. Fetal Bradycardia

It refers to a baseline value below 110 bpm lasting for more than 10 minutes (Fig. 21.6).

Pathophysiology: This is due to initial chemoreceptor activation in response to fetal hypoxia leading to fetal autonomic system activation.

2. Fetal Tachycardia

It refers to a baseline value above 160 bpm lasting more than 10 minutes (Fig. 21.7).

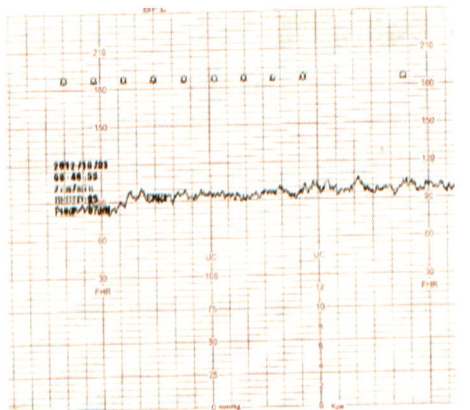

Fig. 21.6: Fetal bradycardia with baseline of 90 bpm.

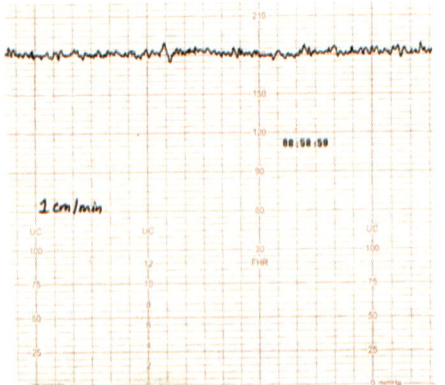

Fig. 21.7: Tachycardia with baseline of 180 bpm.

Pathophysiology: In the initial phase of a non-acute fetal hypoxemia, catecholamine secretion may lead to tachycardia.

3. Reduced Variability

This refers to a bandwidth amplitude below 5 bpm for more than 50 minutes in baseline segments or for more than 3 minutes during decelerations[4] (Figs 21.8A to D; paper speed 1 cm/min).

Pathophysiology: This reduced variability is due to Central Nervous System hypoxia or acidosis leading to reduced autonomic drive. But this pattern of CTG is also seen in previous cerebral injury,[5] infection, parasympathetic blockers or CNS depressants and even during deep sleep. In deep sleep the variability is usually towards lower normal range and the bandwidth amplitude is seldom less than 5 bpm. In hypoxia during labour, reduced variability is very unlikely to occur following an initially normal CTG. It is in most of the cases preceded by or associated with concomitant decelerations and a rise in the baseline.

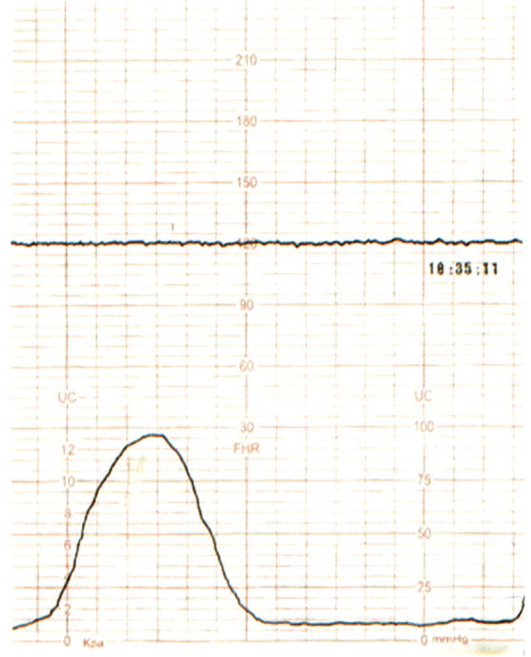

Fig. 218A: Absent variability.

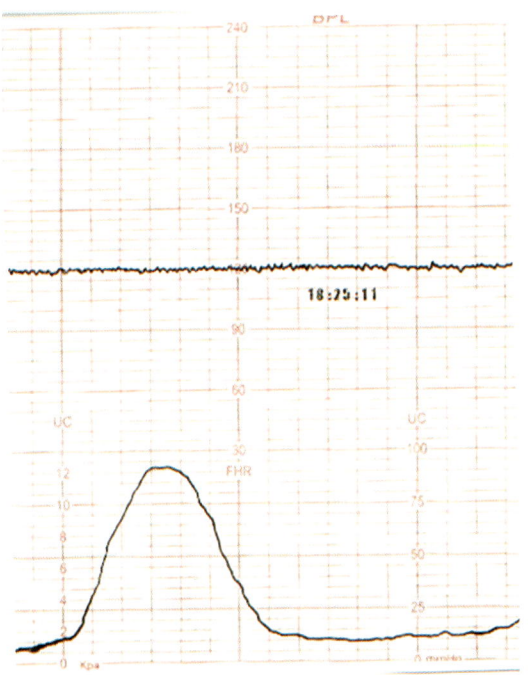

Fig. 21.8B: Minimal variability (<5 bpm).

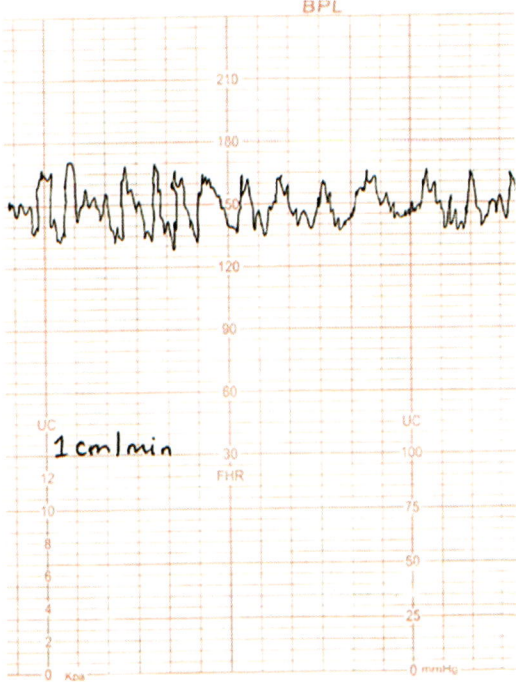

Fig. 218D: Marked variability (>25 bpm).

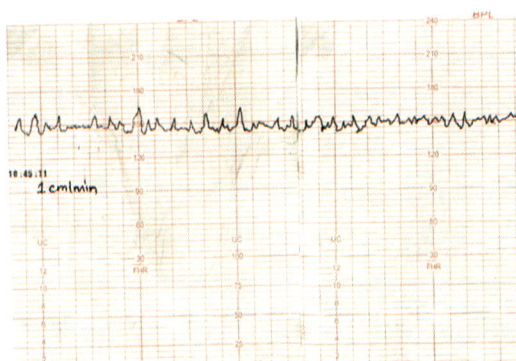

Fig. 218C: Moderate variability (> 5– < 25 bpm).

4. Increased Variability (Saltatory Pattern)

This refers to a bandwidth of more than 25 bpm persisting for more than 30 minutes (Fig. 21.9).

Pathophysiology: Although no clear cut pathophysiology has been understood, this may be associated with repeated decelerations, when hypoxia/acidosis progresses rapidly. It may be assumed to be caused by fetal autonomic instability or hyperactive autonomic system.[6]

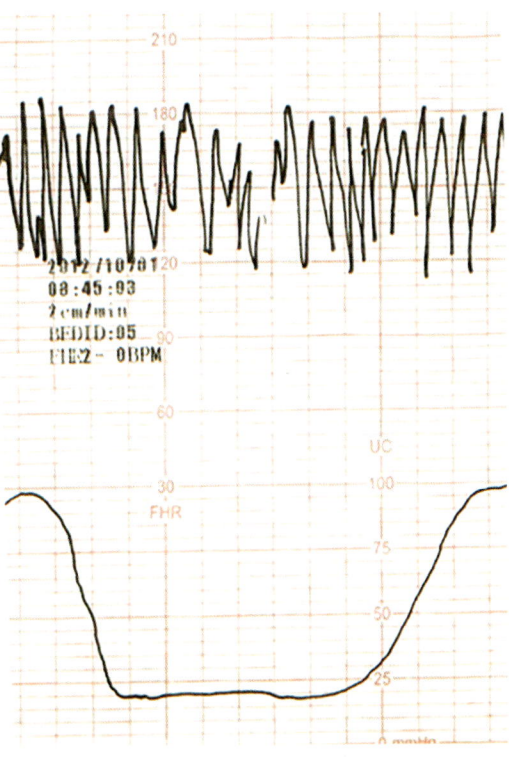

Fig. 21.9: CTG showing increased variability.

5. Absence of Accelerations

Accelerations are abrupt (onset to peak in less than 30 second) rise in fetal heart rates above the baseline of more than 15 bpm for more than 15 second but less than 10 minutes. The absence of accelerations in an otherwise normal intrapartum CTG is of uncertain significance and is unlikely to indicate hypoxia or acidosis.[7]

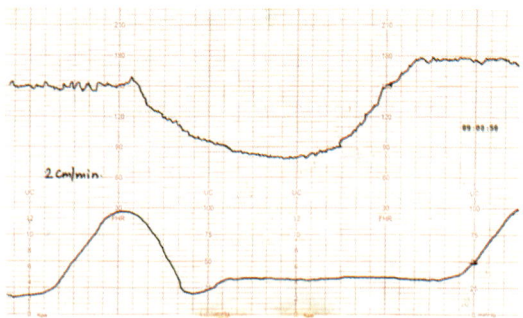

Fig. 21.10: CTG showing late deceleration.

6. Decelerations

This refers to decrease in the FHR below the baseline of more than 15 bpm in amplitude and lasting more than 15 seconds.

6A: Variable decelerations (V-shaped): These are the decelerations which show a rapid drop (onset to nadir in less than 30 seconds), good variability within the deceleration and a rapid recovery to the baseline. They are of varying size and shape and not in a definite relation with the uterine contraction.

Pathophysiology: They are due to baro-receptor mediated response to increased arterial pressure and seen in umbilical cord compression.[8] However these V-shaped decelerations are rarely associated with high degree of hypoxia or acidosis unless evolve to U shaped decelerations with decreased variability within the deceleration and/or with duration exceeding 3 minutes are associated with hypoxia or acidosis.[4,7]

6B: Late decelerations *(U-shaped and/or reduced variability):* These are decelerations with gradual onset and/or gradual return to baseline and/or reduced variability within the deceleration. Gradual onset and return occurs when more than 30 seconds elapse from start/end of the dip and its nadir. These start more than 20 seconds after the contraction onset, nadir after peak, and return to baseline post contraction (FIGO)[7](Fig. 21.10).

Pathophysiology: They are due to chemo-receptor-mediated response to fetal hypoxe-

mia. In presence of a trace without accelerations and reduced variability, the definition of late decelerations also includes those with amplitude of 10–15 bpm.

6C: Prolonged decelerations: These are dips in fetal heart rates persisting for more than 3 minutes.

Pathophysiology: They are also due to chemo-receptor-mediated response to fetal hypoxemia. Decelerations lasting for more than 5 minutes, with FHR maintained less than 80 bpm and decreased variability within the deceleration are frequently associated with acute fetal hypoxia and require an emergent management[4,9] (Fig. 21.11).

7. Sinusoidal Pattern

This refers to a regular, smooth, undulating signal, resembling a sine wave, with

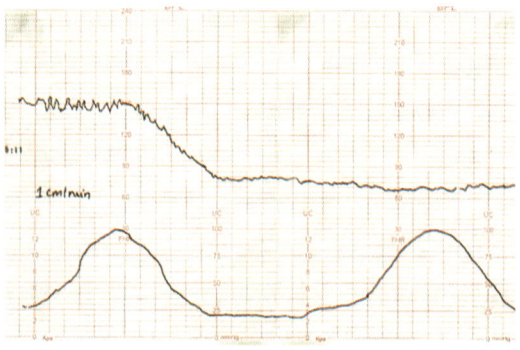

Fig. 21.11: Prolonged deceleration.

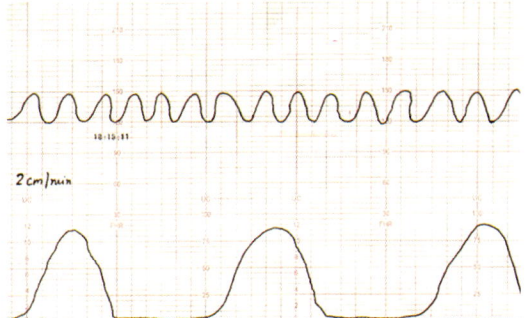

Fig. 21.12: Sinusoidal pattern (Paper speed 1 cm/min).

an amplitude of 5–15 bpm with a frequency of 3–5 cycles per minute. This pattern lasts more than 20 minutes and coincides with absent accelerations[10] (Fig. 21.12).

Pathophysiology: Although not completely understood, this occurs in association with severe fetal anemia as seen in Rh allo-immunisation, twin-twin transfusion syndrome, feto-maternal hemorrhage and rupture vasa previa. Besides acute fetal hypoxia it has also been seen infections, cardiac malformations, hydrocephalus and gastroschisis.[11]

8. Pseudosinusoidal Pattern

This pattern resembles sinusoidal pattern and has a jagged, saw toothed appearance in comparison to smooth sine wave pattern of sinusoidal tracing. Another differentiating feature is its duration which rarely exceeds more than 30 minutes and is preceded and succeeded by normal patterns before and after.

Pathophysiology: Although not clearly understood pathophysiology, this pattern appears after maternal analgesic intake and during fetal mouth movements. The most distinguishing feature is its short duration (less than 30 minutes).

What Does Evidence Says?

Fetus with high probability of having hypoxia or acidosis can be categorized in pathological category (as per FIGO and NICE)[7,12] or in category III tracings (as per ACOG three tiered classification).[10] It has been seen that in a non-anomalous singleton fetus in term labour, category I tracings are observed in 99%, category II in 84% and category III in 0.1% of tracings.[14]

Although described in detail in other Chapters of this book, Table 21.1 recapitulates and compares the CTG categories described by ACOG, FIGO and NICE. It also summarizes their interpretation and clinical management.[7,10,12]

The various characteristics of the of category III/abnormal CTG tracings (ACOG) or Pathological tracings (FIGO, NICE) have been described in Table 21.2. It needs to be understood that though names of the categories are different, but the features are overlapping and the management remains the same.

Clinical Management

In suspected hypoxia or acidosis in CTG tracings, i.e. pathological (NICE, FIGO) or category III tracings (ACOG), urgent action is warranted to avoid adverse neonatal outcome.[13] The underlying cause responsible for the pattern can be identified and the situation controlled.

Excessive uterine activity is the most common cause of fetal hypoxia/acidosis and can be easily detected by tachysystole on CTG tracing or just by clinical palpation of fundus. It can be reversed either by stopping oxytocin or removing prostaglandin if possible or by giving acute tocolysis with beta agonist (like terbutaline)[14] or Atosiban.[15]

Sudden maternal hypotension following epidural analgesia during labour can be reverted by rapid fluid administration.

Aortocaval compression occurring in supine position leads to decreased perfusion of placenta. Moreover, supine posture can also be associated with heightened uterine activity due to stimulation of the sacral plexus by the uterine weight. These cases can be managed

Table 21.1: CTG classification criteria, interpretation and recommended management.

Categorisaation of CTG Tracings			
ACOG	**Category I (Normal)**	**Category II (Indeterminate)**	**Category III (Abnormal)**
FIGO/NICE	**Normal**	**Suspicious**	**Pathological**
Interpretation	Strongly predictive of normal fetal acid–base status	Not predictive of abnormal fetal acid–base status but inadequate evidence to classify either in normal or abnormal/pathological category	Associated with abnormal fetal acid–base at time of observation

Clinical Management	Monitoring in a routine manner (continue CTG, talk to woman and birth companions about what is happening); no specific action	Require evaluation and continued surveillance
		• Use ancillary tests to ensure fetal well-being
		• Action to correct reversible causes if identified (correct maternal posture, stop labour stimulation, treat maternal hypotension)
		• Talk to woman and birth companions about what is happening
			Prompt evaluation and management
			• Obtain senior consultant review
			• Exclude acute events (cord prolapse/abruption/rupture uterus)
			• Correct reversible causes (uterine hyperstimulation or hypotension)
			• Talk to the woman and the her companions about what is happening·
			• If CTG still pathological/abnormal
			  a. use ancillary tests
			  b. expedite delivery
			  c. take woman's preference in account.

just by changing the posture to lateral position which can normalize the CTG changes. Cord compression leading to variable deceleration tracing can be managed by changing the posture of patient or by doing amniofusion.[16]

There is no evidence that administration of intravenous fluids or oxygen to the mother can improve fetal oxygenation and normalize the CTG.[17,18]

CONCLUSION

Limited knowledge of fetal oxygenation pathophysiology, poor CTG interpretation and inadequate clinical management may lead to unnecessary obstetrical intervention and poor pregnancy outcome. Thus sound understanding of cardiotocographic changes in acidosis is needed to diagnose the underlying cause especially in cases of a suspicious or pathological tracing. This is important to judge the reversibility of the underlying condition, to decide about the timing of delivery so as to avoid fetal hypoxia or acidosis and unnecessary obstetric intervention. In case of suspicious CTG pattern, underlying cause should be corrected before the development of pathological tracing. If a pathological tracing ensues, aim should be further evaluation or rapid delivery of the baby, before asphyxia sets in.

Table 21.2: Showing characteristic features of CTG tracings associated with fetal acidosis (hypoxia) as classified by ACOG, NICE and FIGO

	ACOG	NICE	FIGO
	Category III (Abnormal) (any of the # feature or only @ feature)	Pathological tracings (2 non-reassuring features* or 1 abnormal feature**)	Pathological tracings
Baseline	Bradycardia# Sinusoidal@	110–109/161–180* <100/>180/sinusoidal pattern >10 minutes**	<100 bpm
Variability	Absent baseline variability with any of the # feature.	<5 for >40 to <90 minutes* <5 for ≥90 minutes**	a. Reduced variability for >50 minutes or b. Increased variability for >30 minutes or c. Sinusoidal pattern for >30 minutes
Decelerations	Recurrent late decelerations# Recurrent variable decelerations#	a. Early deceleration b. Variable deceleration c. Single prolonged deceleration up to 3 minutes.* d. Atypical variable decelerations e. Late deceleartions f. Single prolonged deceleration >3 minutes**	a. Repetitive late or b. Prolonged during >30 minutes or c. Prolonged >20 minutes with reduced variability or d. One prolonged deceleration >5 minutes

REFERENCES

1. Low JA, Muir DW, Pater EA, Karchmar EJ. The association of intrapartum asphyxia in the mature fetus with newborn behavior. American Journal of Obstetrics and Gynecology 1990;163:1131–5.

2. Diogo Ayres-de-Campos. FIGO Consensus guidelines on intrapartum fetal monitoring: physiology of fetal oxygenation and main goals of intrapartum fetal monitoring. Journal of Gynecol & Obstetrics 2015;131(1):5–8.

3. D. Ayres-de-Campos. Acute fetal hypoxia/acidosis. Obsteric Emergencies. Springer International Publishing Switzerland 2017. doi 10.1007/978-3-319-41656-4_2.

4. Hamilton E, Warrick P, O'Keeffe D. Variable decelerations: do size and shape matter? J Matern Fetal Neonatal Med 2012;25:648–53.

5. Nelson KB, Dambrosia JM, Ting TY, Grether JK. Uncertain value of electronic fetal monitoring in predicting cerebral palsy. N Engl J Med 1996; 334(10):613–8.

6. Nunes I, Ayres-de-Campos D, Kwee A, Rosen KG. Prolonged saltatory fetal heart rate pattern leading to newborn metabolic acidosis. Clin Exp Obstet Gynecol 2014;41(5):507–11.

7. Ayres-de-Campos D, Spong CY, Chandraharan E; FIGO Intrapartum Fetal Monitoring Expert Consensus Panel. FIGO consensus guidelines on intrapartum fetal monitoring: Cardiotocography. See comment in PubMed Commons below Int J Gynaecol Obstet 2015;131(1):13–24.doi: 10.1016/j.ijgo.2015.06.020.

8. Ball RH, Parer JT. The physiologic mechanisms of variable decelerations. Am J Obstet Gynecol 1992;166:1683–9.

9. Cahill AG, Roehl KA, Odibo AO, Macones GA. Association and prediction of neonatal acidemia. Am J Obstet Gynecol 2012;207:206.e1–8.

10. Intrapartum Fetal heart rate monitoring: Nomenclature, Interpretation and general management principles. ACOG Practice Bulletin no. 106, July 2009.

11. Modanlou HD, Murata Y. Sinusoidal fetal heart rate pattern: reappraisal of its definition and clinical significance. J Obstet Gynaecol Res 2004;30:169–80.

12. Intrapartum Care: NICE guideline CG190, February 2017.

13. Jackson M, Holmgren CM, Esplin MS, Henry E, Varner MW. Frequency of fetal heart rate categories and short-term neonatal outcome. Obstet Gynecol 2011;118(4):803–8.

14. Heus R, Mulder EJH, Derks JB, Visser GHA. Acute tocolysis for uterine activity reduction in term labour:a review. Obstet Gynecol Surv 2008;63(6):383–8.

15. Heus R, Mulder EJH, Derks JB, Kurver PHJ, van Wolfswinkel L, Viseer GHA. A prospective randomized control trial of acute tocolysis in term labor with atosiban or ritodrine. Eur J Obstet Gynecol Reprod Biol 2008;139:139–45.

16. Hofmeyr GJ, Lawrie TA. Amniofusion for potential or suspected umbilical cord compression in labour.Cochrane Database Syst Rev 2012;10: CD003401.

17. Simpson KR, James DC. Efficacy of intrauterine resuscitation techniques in improving fetal oxygen status during labor. Obstet Gynecol 2005; 105:1362–8.

18. Fawole G, Hofmeyr GJ. Maternal oxygen administration for fetal distress. Cochrane Database Syst Rev 2012;12:CD000136.

Importance of
Fetal Blood Sampling

Sameer Dikshit

Intrapartum fetal surveillance frequently involves the use of a Cardiotocograph (CTG).[1] The CTG is an electronic method of simultaneously recording Fetal Heart Rate (FHR), fetal movements and uterine contractions to identify the probability of fetal hypoxia.[2] In nearly half of all CTG tracings, an abnormal FHR is observed, but only a small proportion of these fetuses are actually hypoxic.[3] Metabolic acedemia occurs in 2% of all births. Over 90% of these infants will not develop cerebral palsy.[4]

TESTS TO MONITOR FETAL WELL-BEING IN LABOUR

Several ancillary tests for continuous FHR monitoring have been proposed to decrease the false positive rate, or even to replace it completely. These include:

- Fetal scalp blood sampling
- Fetal scalp stimulation
- Fetal ECG ST analysis
- Fetal oximetry.

The gaps in understanding of the precise pathophysiology of the development of metabolic acidosis during labour hinders the efficacy of ALL of the above fetal tests.[5]

Fetal scalp stimulation during vaginal examination is a noninvasive assessment of the fetus that may provide assurance alongside continuous FHR monitoring and fetal scalp blood sampling in cases of suspected fetal compromise. The likelihood ratio of an acceleration following fetal scalp stimulation for having low scalp pH is 0.5.[7] This means that, at low probabilities, the risk of fetal acidosis after an acceleration following stimulation reduces by 50%.

Fetal electrocardiograph (ECG) ST analysis: A meta-analysis of randomised controlled trials has shown fetal ECG ST analysis reduces the need for fetal scalp blood sampling by about 40%. However, the trends to lower rates of low Apgar scores and acidosis were not statistically significant.[8] Currently, RANZCOG states there is insufficient evidence to recommend routine fetal ECG ST analysis for use in intrapartum fetal surveillance.[1]

Fetal pulse oximetry: Currently, RANZCOG states that there is insufficient evidence to recommend routine fetal pulse oximetry for use in intrapartum fetal surveillance.[1] Fetal pulse oximetry results are affected by the presence of meconium and blood and a recent Cochrane review has concluded that fetal pulse oximetry is not associated with improvement in fetal outcomes.[6]

PHYSIOLOGY OF FETAL AND PLACENTAL CIRCULATION

For understanding the fetal blood sampling results, understanding fetal and placental circulations is essential. Let us briefly go over the physiology.

The fetal heart pumps deoxygenated blood to the placenta via the two umbilical arteries. At the placenta there is close contact between the fetal and maternal blood without actual mixing. This allows for free exchange of the gases. The placenta in fact works as a lung for the baby *in utero*. Oxygenated blood is pumped back to the fetus through umbilical vein. The free gas exchange depends on normal flow of blood on maternal and fetal sides.

Aerobic Metabolism

Normally, oxygen gets transferred to the fetal tissues where it is used to produce energy from glucose. Fetal blood has more concentration of hemoglobin which has more affinity for oxygen. This ensures that adequate oxygen is available for the fetal tissues. The measure of amount of oxygen carried is "Partial Pressure of oxygen"—pO_2.

In tissues, oxygen is utilised to produce energy from glucose. Carbon dioxide is a by-product of this process. It reacts with water to form H^+ and HCO_3^- ions. H^+ ions are buffered by Hb in fetal RBCs while HCO_3^- passes into extracellular fluid. Free H^+ ions in the fetal blood are represented as acidic pH of the blood.

Fetal blood can carry a lot of H^+ ions due to the presence of buffers which neutralise the acidity.

The events are reversed when blood reaches placenta. CO_2 is transferred to the placenta where it is eliminated. This causes more H^+ to be dissociated from fetal Hb. It combines with HCO_3^- to form CO_2 which is eliminated at placenta. In this manner CO_2 is eliminated from fetal blood, H^+ ions are disposed of and the pH of the fetal blood is restored.

The fetal blood flowing through umbilical arteries to the placenta has low pO_2, high pCO_2, high concentration of H^+ ions and acidic pH. While, the blood flowing away from the placenta through umbilical vein has high pO_2, low pCO_2, low concentration of H^+ ions and alkaline pH.

Under normal circumstances, fetus satisfies most of its energy need by aerobic metabolism.

This ideal situation is altered when the gas exchange in placenta is interrupted. This happens in case of:

- Maternal hypotension
- During excessive uterine contractions
- Placental abruption
- Altered uteroplacental circulation in IUGR/PIH
- Cord compression
- Fetal anemia.

This results in reduced supply of O_2 to fetal tissues and accumulation of CO_2 in fetal blood. Accumulation of CO_2 in blood precludes clearance of H^+ from fetal blood. As a result, H^+ ions are retained in blood causing fall in pH making it acidotic. Thus, the blood returning to the fetus through umbilical vein has high pCO_2 and low pH. This is called **Respiratory Acidemia**. When this persists for a long-time, the fetal tissue has low pH.

The fetus is well-adapted to survive in conditions of low oxygen tension for brief time intervals. This happens due to increased affinity of Fetal Hb for O_2. The fetal tissues normally have excess supply of oxygen and have some reserve supply.

If, however, these conditions persist for longer time, then fetus makes some adjustments to satisfy its energy needs.

It redistributes blood in favour of vital organs and switches to anaerobic metabolism.

Anaerobic Metabolism

Anaerobic metabolism leads to glycogen stores being used to produce energy with lactic acid as a by-product. This leads to fall in pH of the blood. This is called **Metabolic Acidemia**.

When the tissues finally are faced with acidic condition then this condition is called **Acidosis**.

The metabolic and respiratory types of acidosis can be distinguished by measuring **Base Deficit**. The base deficit value is large in metabolic acidosis than in respiratory type.

Anaerobic metabolism not only causes acidosis but it also is very inefficient way of producing energy. It breaks down carbohydrate stores 19 times more quickly than aerobic metabolism. The reserves are used up quickly and the fetus faces hypoxia. Thus anaerobic metabolism leads to hypoxia and acidosis. Both these are toxic to the fetal tissue.

FETAL SCALP BLOOD SAMPLING

A recent systematic review of intermittent auscultation versus continuous cardiotocograph in both low and high-risk women reveals a significant increase in the cesarean section rate, whether fetal blood sampling was deployed in labour or not.[1,2,4] It is therefore possible that the availability of fetal blood sampling in labour will lessen the increase in the cesarean rate that comes as a consequence of using continuous CTG.[1]

Although fetal scalp blood sampling is generally considered to be a safe test, rare complications (e.g. hemorrhage, scalp abscess and drainage of cerebrospinal fluid) and questions regarding the accuracy of current normal and abnormal values for fetal scalp pH (derived from two small studies) as well as the accuracy of pH levels obtained from a fetal scalp venous sample have led some medical experts to question if this procedure is clinically and scientifically acceptable.[5,9] RANZCOG supports the practice of fetal scalp blood sampling, particularly in larger units that have ready access to operative delivery if required. However, RANZCOG acknowledges that it is not practical for all hospitals to provide fetal blood sampling. For example, in some hospitals, undertaking fetal blood sampling may delay a necessary delivery and thereby worsen outcomes by lengthening the decision to delivery interval for an emergency cesarean section.[1] In the past, some hospitals interested in providing fetal blood sampling were unable to because of the costs of maintaining the necessary hardware. More recently, the introduction and validation of scalp lactate measurement has provided an affordable alternative.[1] If fetal scalp blood sampling is indicated, the use of scalp lactate rather than pH measurement will provide an easier and more affordable adjunct to electronic FHR monitoring for most units.[1,10] If fetal scalp blood sampling is performed, the scalp lactate or pH result should be interpreted taking into account any previous lactate or pH measurement, the rate of progress in labour and the clinical features of the woman and baby.[1,2] In situations where fetal blood sampling is contraindicated (see below) or not possible, decisions regarding delivery should take into account the severity of the FHR abnormality and the clinical situation.

Delivery should be expedited where:

- There is clear evidence of serious fetal compromise (fetal scalp blood sampling should not be undertaken
- Cardiotocograph abnormalities are of a degree requiring further assessment, but fetal scalp blood sampling is contraindicated, clinically inappropriate or unobtainable
- The decision to delivery interval may be prolonged by virtue of location, clinical staff availability, patient factors or access to clinical services.

Fetal Scalp Lactate and pH Levels

A randomised, controlled multicentre trial showed pH analysis and lactate analysis of fetal blood have comparable results in the management of intrapartum fetal compromise.[3]

- The average fetal scalp blood pH is 7.33 in normal labour

- pH > 7.25 and lactate < 4.1 mmol/L is considered normal
- The mean umbilical artery pH after uncomplicated pregnancy and labour ranges from 7.25 to 7.31 in different studies[13]
- Thresholds for lactate may vary between institutions. Institutions should have local guidelines for lactate thresholds.

Indications for Fetal Scalp Blood Sampling

Factors including clinical history, parity, evolution of the FHR pattern, stage and rate of progress in labour influence the decision for fetal scalp blood sampling. Fetal scalp blood estimation may be of value in the following circumstances:

- Bradycardia
- Complicated tachycardia
- Recurrent decelerations
- Prolonged episodes of bradycardia or undefined deceleration patterns
- Prolonged loss of variability which does not spontaneously correct with fetal stimulation
- Miscellaneous, e.g. nonspecific concerns about fetal well-being.

Contraindications for Fetal Scalp Blood Sampling

- Clear evidence on cardiotocograph of serious, sustained fetal compromise
- Maternal infection, e.g. Hepatitis B, C, HIV, herpes simplex virus and suspected intrauterine sepsis
- Fetal bleeding disorders (e.g. suspected fetal thrombocytopenia, hemophilia)
- Face or brow presentation
- Fetal blood sampling is not generally recommended in pregnancies at less than 34 + 6 weeks of gestation because delivery may be inappropriately delayed in a small "at risk" fetus that may sustain damage earlier than would be expected in a term fetus[1]

- If a fetus is in a breech presentation during labour and is exhibiting signs of fetal compromise that are not readily remediable, it would be more appropriate to deliver the baby by cesarean section than to undertake fetal blood sampling.

MANAGEMENT OF FETAL SCALP BLOOD SAMPLING

In tertiary centres, fetal scalp blood sampling should be considered part of routine care for the management team when indicated, and a competent resident medical officer or registrar should be able to fulfil.

Explain the following to the woman:[2]

- Why the test is being advised
- The blood sample will be used to measure the level of acid in the baby's blood, to see how well the baby is coping with labour
- The procedure will require her to have a vaginal examination using a small device similar to a speculum
- A sample of blood will be taken from the baby's head by making a small scratch on the baby's scalp
- This will heal quickly after birth, but there is a small risk of infection
- The procedure can help to reduce the need for further, more serious interventions
- What the different outcomes of the test may be (normal, borderline and abnormal) and the actions that will follow each result
- There is a small chance that it will not be possible to obtain a blood sample (especially if the cervix is less than 4 cm dilated)
- If a sample cannot be obtained, a cesarean section or instrumental birth (forceps or ventouse) may be needed because otherwise it is not possible to find out how well the baby is coping
- Do not carry-out fetal blood sampling if any contraindications are present,

including risk of maternal-to-fetal transmission of infection or risk of fetal bleeding disorders.

Procedure for Fetal Scalp Blood Sampling

Position

The preferred maternal position is left-lateral position with hips well-flexed and the lower leg extended. The upper leg should be flexed (held by an assistant or positioned in a stirrup) with the buttocks extending over the edge of the bed to allow the clinician to be positioned below the level of the maternal vagina. If lithotomy position is used, ensure a lateral wedge is used to prevent aortocaval compression.

Procedure

Attach the fetal scalp blade (depth of 2 mm) to an introducer. Under direct vision, insert amnioscope with light source into the posterior fornix. The clinician obtaining the scalp sample should aim to angle the amnioscope downward below the horizontal plane. Once past the anterior lip of the cervix, angle the cone anteriorly into the cervix to visualise the presenting part. Clean the fetal scalp surface with chlorhexidine/alcohol-soaked gauze. Apply sterile liquid paraffin to the fetal scalp (forms a non wettable surface and encourages beading of fetal scalp blood). Make a quick stab with the fetal scalp blade/introducer to achieve a clean incision on the fetal scalp. As the fetal blood appears, insert the heparinised capillary tube to touch the drop of blood, and keeping the tube angled downward, the blood is allowed to flow by gravity.

FBS for pH: Let the tube fill with at least 2 cm of blood (without air bubbles or liquor).

FBS for lactate: A minimum of 5 microlitres of blood is required (without air bubbles or liquor).

Immediately pass the sample to an assistant for processing. Obtain two samples. Apply pressure with a swab to the fetal scalp over the next two contractions and observe to ensure the bleeding has stopped.

Results: Table 22.1.

Table 22.1: Fetal blood sampling	
Lactate and pH results	**Management**
Lactate < 4.1 mmol/L pH ≥ 7.25	**Fetal scalp blood sample is normal** Offer repeat sampling no more than 1 hour later if this is still indicated by the cardiotocograph trace, or sooner if additional signs of fetal compromise or abnormal features become evident
Lactate 4.1– 4.7 mmol/L pH 7.21–7.24	**Fetal blood sample result is borderline** Offer repeat sampling no more than 30 minutes later if this is still indicated by the cardiotocograph trace, or sooner if additional signs of fetal compromise or abnormal features become evident.[2] Take into account the time needed to take a fetal blood sample when planning repeat fetal sampling.[2]
Second fetal scalp blood sample result is stable and no further signs of fetal compromise	If the cardiotocograph trace remains unchanged and the fetal scalp blood sample result is stable (that is lactate or pH is unchanged) after a second test, further samples may be deferred unless additional signs of fetal compromise or abnormal features are seen.[2]
Lactate > 4.7 mmol/L pH ≤ 7.20	Delivery indicated. Rapid deterioration in features of fetal compromise requires obstetric review of timing and mode of delivery. Consider the woman's complete history (e.g. presence of meconium, progress, fetal scalp lactate or pH value) when assessing need for cesarean section (category 1)
Lactate of ≥ 5.8 mmol/L pH <7.00	Requires an urgent assisted vaginal delivery if possible or a category 1 cesarean section.

Fetal Scalp Blood Sample cannot be Obtained

If a fetal scalp blood sample is indicated and the sample cannot be obtained, but the associated scalp stimulation results in FHR accelerations, a decision whether to continue the labour or expedite the birth will be made in consideration of the clinical circumstances and in discussion with the obstetrician on call and the woman.[2] If a fetal scalp blood sample is indicated but a sample cannot be obtained and there is no improvement in the cardiotocograph trace, advise the woman that the birth should be expedited.[4] Notify anesthetist and pediatrician. Urgency of delivery should take into account the severity of fetal compromise and relevant maternal factors.

No sample or one contaminated with liquor or inadequate volume sample obtained— Review indication for fetal scalp blood sampling and current cardiotocograph trace. Consider need for delivery in consultation with obstetrician on call.

UMBILICAL CORD BLOOD GASES

Measures to assess a newborn baby:
- Knowledge of antepartum and intrapartum events
- Apgar score
- Need for resuscitation
- Neonatal behaviour
- Cord blood acid base result.

Umbilical cord blood gases assessment gives valuable insight. Cord blood assessment gives valuable information about:

- How well the oxygen has been maintained to the fetus during labour
- How well the fetus has eliminated the waste product CO_2
- Unlike APGARs, cord blood analysis is objective
- It gives us an idea about placental gas exchange in labour

- It gives useful insight about planning neonatal management.

Importance of Umbilical Cord Blood Sampling[17]

Newborn Assessment

Apgar score identifies those babies who need resuscitation at birth and helps to direct early neonatal care. The longer an Apgar score remains low, the more likely it is to reflect a significant problem. Cord blood assessment distinguishes between low APGAR score due to events occurring just before the delivery. These do not have any long-term consequence for the fetus. Let us take example of two fetuses. The first fetus had IUGR in antenatal period and is born with APGAR 8/10. The second fetus had uneventful antenatal history but had supine hypotension during LSCS. It is born with APGAR 6/10. The first fetus has more risks than the second. The difference between these two cases can be made out by cord blood gas analysis. Large difference between arterial and venous pH values suggests acute event causing low APGAR values.

Neonatal Management

If a baby has normal APGAR at birth but shows abnormal cord blood gases, then the baby should be watched in neonatal period for complication like hypoglycemia.

Audit

Use of cord blood gases to correlate with CTG abnormalities helps obstetricians to understand fetal heart changes better. It has been shown that the practice of using cord blood values in perinatal meetings helps to improve the skills of CTG interpretation.

Medicolegal Purpose

Cord blood gas analysis can never be a substitute for other measures. It only tells us if a baby has been deprived of oxygen during labour but it cannot tell us if the baby has suffered harm as a result.

Fetal arterial and venous cord gases (pH and base excess) are not required for uncomplicated term spontaneous vaginal births.

Indications for Obtaining Umbilical Cord Blood Sample

Obtain fetal arterial and venous cord gases (pH and base excess) at time of birth where:

- Fetal scalp bloods have been taken
- Operative vaginal delivery or cesarean section is required
- Baby is less than 37 + 0 weeks of gestation
- Multiple pregnancy
- Breech vaginal birth
  - Baby's condition is poor at birth
  - Meconium stained liquor is present
- It is generally accepted that arterial and venous pH should differ by 0.03 to be sure that the artery has been sampled.[13]

COLLECTING CORD BLOOD AFTER DELIVERY

- The cord should be clamped as soon as possible
- The segment of the cord should be isolated between two sets of clamps
- The cord segment can be kept at room temperature for about half an hour
- If the delay is going to be longer, then it needs to be stored in refrigerator
- Sample needs to be collected from both the vein and artery
- When the blood is collected in syringes, they have to be heparinised
- A large difference in pH between arterial and venous blood it suggests an acute problem in second stage
- If the difference is small, then it suggests a longstanding problem

- Arterial pH <7.05 and venous pH <7.10 are clinically important
- Arterial base deficit >12 mmol/l and venous base deficit >10 mmol/l are clinically important
- Acidemia + significant base deficit suggests metabolic acidemia.

REFERENCES

1. The Royal Australian and New Zealand College of Obstetricians & Gynaecologists (RANZCOG): Intrapartum Fetal Surveillance. Clinical Guidelines–Third Edition; 2014 (Level IV). Available at URL:http://www.ranzcog.edu.au
2. National institute for health and care excellence (NICE). Intrapartum care: Care of healthy women and their babies during childbirth. NICE clinical guideline 190;2014 (Level IV). Available from URL: http://www.nice.org.uk/guidance/cg190.
3. Wiberg-Itzel E, Lipponer C, Norman M, Herbst A, Prebensen D, Hansson A, Bryngelsson AL, Christoffersson M, Sennström M, Wennerholm UB, Nordström L. Determination of pH or lactate in fetal scalp blood in management of intrapartum fetal distress: randomised controlled multicentre trial BMJ 2008;336:1284–7 (Level II).
4. Royal College of Obstetricians & Gynaecologists (RCOG). The Use of Electronic Fetal Monitoring, Evidence-based Clinical Guideline Number 8. RCOG Clinical Effectiveness Support Unit, London: RCOG Press; 2001.
5. Chandraharan E. Fetal scalp blood sampling during labour: is it a useful diagnostic test or a historical test that no longer has a place in modern clinical obstetrics? BJOG 2014;121:1056–62.
6. East CE, Begg L, Colditz PB, Lau R. Fetal pulse oximetry for fetal assessment in labour. Cochrane Database of Systematic Reviews 2014, Issue 10. Art. No.: CD004075. doi: 10.1002/14651858. CD004075.pub4. Available from URL: http://onlinelibrary.wiley.com/doi/10.1002/14651858. CD004075.pub4/pdf/standard.
7. Skupski DW, Rosenberg CR, Eglinton GS. Intrapartum fetal stimulation tests: A meta-analysis. Obstet Gynecol 2002;99:129–34.
8. Steer PJ, Hvidman LE. Scientific and clinical evidence for the use of fetal ECG ST segment analysis (STAN). Acta Obstet Gynecol Scand 2014;93:533–8.

9. Alfirevic Z, Devane D, Gyte GML. Continuous cardiotocography (CTG) as a form of electronic fetal monitoring (EFM) for fetal assessment during labour. Cochrane Database of Systematic Reviews 2013, Issue 5. Art. No.: CD006066. doi: 10.1002/14651858.CD006066.pub2. Available from URL: http://onlinelibrary.wiley.com/doi/10.1002/14651858.CD006066.pub2/pdf/standard.

10. East CE, Leader LR, Sheehan P, Henshall NE, Colditz PB, Lau R. Intrapartum fetal scalp lactate sampling for fetal assessment in the presence of a non-reassuring fetal heart rate trace. Cochrane Database of Systematic Reviews 2015, Issue 5. Art. No.: CD006174. DOI: 10.1002/14651858. CD006174.pub3. Available from URL: http://onlinelibrary.wiley.com/doi/10.1002/14651858. CD006174.pub3/pdf/standard.

11. Andres RL, Saade G, Gilstrap LC, Wilkins I, Witlin A, Zlatnik F, et al. Association between umbilical blood gas parameters and neonatal morbidity and death in neonates with pathologic fetal acidemia. Am J Obstet Gynecol 1999;181(4): 867–71 (Level IV).

12. Freeman RK, Garite TJ, Nageotte MP. Fetal Heart Rate Monitoring. 3rd ed. Chapter 8. Philadelphia: Lippincott Williams & Wilkins 2003.

13. Vandenbussche FPHA, Oepkes D, Keirse MJNC. The merit of routine cord blood pH measurement at birth. J Perinat Med 1999;27:158–65 (Level IV).

14. Royal College of Obstetricians and Gynaecologists (RCOG). Clinical risk management for Obstetricians and Gynaecologists. Clinical governance Advice No 2. October 2005. Available at URL:http://www.rcog.org.uk/resources/Public/pdf/improvingpatientsafety_cga.pdf.

15. SA Health. Standards for the Management of Category One Caesarean Section in South Australia. Government of South Australia, Department of Health. December 2011.

16. The Royal Australian and New Zealand College of Obstetricians and Gynaecologists (RANZCOG): Categorisation of urgency for caesarean section. C-Obs 14. July 2012. Available from URL: http://www.ranzcog.edu.au

17. K2MS™ Perinatal Training Programme (PTP) https://training.k2ms.com.

Medicolegal Considerations of CTG

Vikram Sinai Talaulikar

The principle aim of intrapartum fetal monitoring is to detect changes in the Fetal Heart Rate (FHR) that suggest a possibility of fetal hypoxia and metabolic acidosis so that timely action can be taken to prevent adverse outcomes. Continuous electronic fetal heart rate monitoring (EFM) by Cardiotocography (CTG) has formed the mainstay of fetal surveillance in high-risk pregnancies in most of the developed world. However, obstetric malpractice claims and their escalating costs, together with the increase of medical malpractice insurance premiums, have become a major concern for maternity service providers.[1] Obstetric claims related to mishaps during labour, leading to intrapartum hypoxia and long-term neurological damage to the newborn, are associated with enormous costs of financial compensation for families who have to suffer the emotional distress and care for a possibly life-long handicapped child. Many medicolegal cases have similar underlying problems such as:

a. Inability to interpret CTG trace
b. Inappropriate action on an abnormal trace
c. Technical errors
d. Documentation issues.

Steps can be taken to reduce these risks by regular CTG training for labour ward staff, risk management and provision of support as well as supervision to healthcare staff when necessary.

OBSTETRIC LITIGATIONS—SCALE OF THE PROBLEM

About one out of every ten newborns who later develop cerebral palsy has evidence of isolated intrapartum hypoxia as a cause. Despite its low specificity for hypoxia, the CTG continues to be the central documentary evidence for all claims for fetal asphyxia.[1,2] The speed of onset and progression of hypoxia can be predicted by distinct types of CTG. It is also possible to predict the type of neurological injury based on the type of CTG that gives rise to hypoxia.

Although in some of the cases, interventions may have prevented or decreased the severity of cerebral palsy, it has been shown overall that FHR patterns are poor predictors of cerebral palsy.[1–4] The low specificity of CTG for fetal hypoxia often necessitates secondary or definitive tests to confirm fetal acid-base status in labour, such as fetal blood sampling, vibro-acoustic stimulation, fetal oximetry or ST segment analysis of fetal ECG.

A report from United Kingdom in 2003 found that litigation arising from maternity services accounted for 60–70% of the total sum paid out by the National Health Services Litigation Authority (NHSLA), but comprised only 26% of the workload in England.[5] Litigation can be an unpleasant experience for the healthcare staff involved. Malpractice claims can have significant long-term consequences for the working lives of midwives or obstetricians.

CHALLENGES OF INTRAPARTUM FETAL MONITORING— ROLE OF CTG

Despite its shortcomings continuous intrapartum CTG remains the predominant method of intrapartum fetal surveillance wherever facilities allow, mainly because of medicolegal reasons (it provides a graphical trace record of the FHR throughout labour), also because it is helpful in identifying asphyxiating conditions during labour and because there is no other better independent monitoring modality yet established for widespread clinical use.

When intrapartum CTG was introduced in clinical practice, it was hoped that this method would reduce the incidence of cerebral palsy and mental retardation by 50%. However, this dream was not realised. This may have been because (a) in many cases the asphyxial damage may have begun before labour and (b) acute asphyxia associated with events, such as prolapsed cord, ruptured uterus, ruptured vasa previa or abruption does not always allow sufficient time for intervention before damage is done.[3,6] Also, extreme prematurity and fetal infection could be other variables involved in causation of cerebral palsy.

One large analysis of initial studies concluded that intrapartum fetal death was significantly less common in patients who were monitored with EFM than in those who had auscultation without one to one care.[7] However, subsequent Randomised Controlled Trials (RCTs) comparing EFM with intensive one to one auscultation in term patients found no differences with respect to perinatal mortality, Apgar scores or neonatal intensive care unit admissions.

MEDICOLEGAL ASPECTS OF INTRAPARTUM FETAL MONITORING

Despite its low specificity for hypoxia, the CTG continues to be the central documentary evidence for claims for fetal asphyxia[1]. Other indicators including a low Apgar score at birth have been shown to be subjective and poor predictors of long-term neurological outcomes. Various criteria have been developed by different national bodies to define the term 'birth asphyxia'. The American College of Obstetricians and Gynecologists (ACOG) issued the following in 1992[8]—for perinatal asphyxia to be linked to a neurological deficit in the child, all of the following criteria must be present: (1) profound umbilical artery metabolic or mixed acidemia (pH <7.00), (2) persistence of an Apgar score of 0–3 for longer than 5 minutes, (3) neonatal neurological sequelae (e.g. seizures, coma, hypotonia), and (4) multiorgan system dysfunction (e.g. cardiovascular, gastrointestinal, hematologic, pulmonary, renal). In 1995 the Task Force on Cerebral Palsy and Neonatal Asphyxia of the Society of Obstetricians and Gynaecologists of Canada issued a policy statement in which they stated the same criteria with addition of umbilical artery base deficit of >16 mmol/L.[9]

An international consensus statement subsequently developed in 1999 redefined the following essential and/or additional criteria for birth asphyxia (intrapartum event leading to cerebral palsy).[10]

Essential Criteria

- Profound umbilical artery metabolic acidemia (pH <7.0 and base deficit (BD) >12)
- Early onset of severe or moderate encephalopathy in infants >34 weeks
- Cerebral palsy of a spastic quadriplegic or dyskinetic type
- Exclusion of other identifiable etiologies, such as trauma, coagulation disorders, infectious conditions or genetic disorders. (The 4th criterion was added by ACOG in 2003).[11]

Additional Criteria

- A sentinel hypoxic event occurring immediately before or during labour

- A sudden rapid sustained deterioration in FHR pattern
- Apgar scores of <7 after 5 minutes
- Early evidence of multi-organ ischemic injury
- Early imaging evidence of acute cerebral involvement.

Umbilical cord arterial blood gas analysis at birth has emerged as an important method, used to support or refute a diagnosis of intra-partum asphyxia. Most maternity units now routinely determine umbilical cord arterial and venous blood acid–base status on deliveries where there has been any concern during labour, e.g. operative deliveries, cases where a fetal scalp blood sample was done, pathological CTG trace, those with meconium stained amniotic fluid, bleeding in labour, preterm infants, multiple gestations, vaginal breech deliveries and the depressed infant at birth.

CONCEPT OF CAUSATION

The legal process of the claim involves establishment of liability, causation and quantum. Once a claim is intimated, it is for the claimant to establish that there has been breach of duty, that is standard of care fell below what would be expected and the clinician did not act in accordance with what would be considered appropriate by a responsible body of medical opinion. In birth injury claims, the claimant should prove that, on the basis of the CTG trace, delivery should have been earlier than it was and that this would have prevented some, if not all, of the injury. If the CTG was normal, or the cause of neonatal encephalopathy remains unclear, then a pediatric neurologist's opinion is sought to exclude a host of metabolic, genetic and infectious causes of neurological impair-ment. Some of the questions which may arise during establishment of causation will relate to timing, type and severity of the injury. It is not always possible based on the CTG to accurately determine the timing of injury.

Similarly, a grossly abnormal CTG may suggest possible injury but may not predict severity. The type of injury will often be determined by correlation of clinical picture, trace and imaging findings by the neurologist and/or neuro-radiologist.

HYPOXIC ISCHEMIC ENCEPHALOPATHY (HIE)—GRADING, IMAGING AND DIAGNOSIS

Reduced cardiac output in the setting of hypoxia is referred to as Hypoxia-Ischemia (HI). If an episode of HI is severe enough to damage the brain, it leads within 12 to 36 hours to a neonatal encephalopathy known as Hypoxic Ischemic Encephalopathy (HIE). This clinical syndrome includes seizures, epileptic activity on Electroencephalogram (EEG), hypotonia, poor feeding, and a depressed level of consciousness that typically lasts from 7–14 days.[12] HIE can vary in severity from grade 1 (mild) to grade 3 (severe). Full term infants who suffer from Grade 2 or 3 encephalopathy are known to have a higher risk of long-term neurologic damage. Neuro-radiologist and pediatric neurologist play a key role in determining the causation in such cases (obstetrician/midwife work along but more for determining liabi-lity). Imaging modalities such as neonatal Magnetic Resonance Imaging (MRI) and ultrasonography are used.[13]

The nature of asphyxia can determine the type of brain injury and the neurological outcome as described by Myers in 1975:[14]

- Total asphyxia causes damage to the brainstem and thalamus (athetoid or dyskinetic cerebral palsy)
- Prolonged hypoxia with acidosis causes brain swelling and cortical necrosis (spastic quadriplegic cerebral palsy)
- Prolonged hypoxia without acidosis causes white matter damage
- Total asphyxia preceded by prolonged hypoxia with mixed acidosis causes damage to the cortex, thalamus and basal ganglia.

Acute hypoxia is encountered in situations such as prolonged deceleration or bradycardia while subacute hypoxia is seen in cases with prolonged decelerations where the FHR spends more time below the baseline rate (>90 seconds) and shorter duration at the baseline rate (<30 seconds). These two patterns usually present with acute clinical events or in late first or second stage of labour.

In cases of intrapartum events leading to severe hypoxic injury, the liability will need to be established. Quantum is considered at a later stage to determine what facilities and support are required because of the injury. Liability is usually judged on what a reasonably competent practitioner would have done (the Bolam test). According to the Bolam principle—The test is the standard of the ordinary skilled man exercising and professing to have the specialist skill. A man need not possess the highest expert skill to be at risk of being found negligent. If a breach in the standard of care is established then compensation will be awarded. The Bolitho principle is increasingly being used, especially in cases in which there is an unresolved dispute among medical experts as to whether liability is due to omission or commission of care by the practitioner. According to the Bolitho principle: If it can be demonstrated that the professional opinion is not capable of with standing logical analysis, the judge is entitled to hold that the body of opinion is not reasonable or responsible.

OBSTETRIC LITIGATIONS—REASONS AND COMMON PITFALLS

Confidential inquiries into perinatal deaths and cases of litigation suggest the following key points for litigation:

- Failure to incorporate the overall clinical situation
- Inability to interpret the CTG correctly
- Technical difficulties
- Delay in acting on abnormal CTG
- Poor communication and record keeping.

AVOIDANCE OF LITIGATION AND MANAGEMENT OF ABNORMAL CTG

High-risk pregnancy situations demand special attention during fetal monitoring, as such fetuses have an increased risk of developing hypoxia or acidosis with a deteriorating CTG trace compared to an appropriately grown fetus with clear amniotic fluid.

Following obstetric conditions should alert the obstetric staff to remain highly vigilant during intrapartum fetal monitoring:

- Post-term pregnancy
- Pre-term pregnancy
- Intrauterine fetal growth restriction
- Meconium stained or scanty amniotic fluid
- Intrauterine infection
- Intrapartum bleeding
- Augmented labour (oxytocin)
- Difficult instrumental delivery
- Macrosomia
- Abnormal CTG findings at admission.

Systematic interpretation of CTG and appropriate documentation of suggested action based on overall clinical impression in patient's medical records are critical actions which ensure a high standard of clinical practice as well as go a long way in preventing obstetric litigation.

The National Institute for Health and Care Excellence (NICE) guidelines in United Kingdom recommend continuous cardiotocography in labour if any of the risk factors is identified on initial assessment or arise during labour and it must be explained to the woman why this is necessary.[15]

These risk factors may be:

- maternal pulse over 120 beats/minute on 2 occasions 30 minutes apart
- temperature of 38°C or above on a single reading, or 37.5°C or above on 2 consecutive occasions 1 hour apart
- suspected chorioamnionitis or sepsis
- pain reported by the woman that differs from the pain normally associated with contractions
- the presence of significant meconium

- fresh vaginal bleeding that develops in labour
- severe hypertension: A single reading of either systolic blood pressure of 160 mm Hg or more or diastolic blood pressure of 110 mm Hg or more, measured between contractions
- hypertension: Either systolic blood pressure of 140 mm Hg or more or diastolic blood pressure of 90 mm Hg or more on 2 consecutive readings taken 30 minutes apart, measured between contractions
- a reading of 2+ of protein on urinalysis and a single reading of either raised systolic blood pressure (140 mm Hg or more) or raised diastolic blood pressure (90 mm Hg or more)
- confirmed delay in the first or second stage of labour
- contractions that last longer than 60 seconds (hypertonus), or more than 5 contractions in 10 minutes (tachysystole)
- oxytocin use.

If continuous cardiotocography is needed, it should be explained to the woman that it will restrict her mobility, particularly if conventional monitoring is used. The woman should be encouraged to be as mobile as possible and to change position as often as she wishes. It should be born in mind that it is not possible to categorise or interpret every CTG trace; senior obstetric input is important in these cases.[15]

Cardiotocography is also recommended if intermittent auscultation indicates possible FHR abnormalities. The cardiotocograph can be discontinued if the trace is normal after 20 minutes.

Whenever CTG is employed as a form of fetal monitoring at the beginning of labour, the woman should be offered an explanation that a normal trace is reassuring and indicates that the baby is coping well with labour, but if the trace is not normal there is less certainty about the condition of the baby and further continuous monitoring should be advised.

When interpreting a CTG trace, it is important to follow a methodical approach (Table 23.1).[16] The use of vague descriptors such as 'sleepy trace', 'bad CTG' or 'fetal-distress' should be avoided. NICE evidence-based clinical guidelines (Tables 23.2 to 23.4) promote a systematic approach to CTG interpretation, with individual features classified as reassuring, non-reassuring or abnormal and the overall CTG as normal, suspicious and abnormal.[15] Continuous EFM should be systematically assessed at least once an hour in labour and more often if indicated. Some of the critical issues which need to be considered while recording or evaluating a CTG trace are:

- Patient identity
- Maternal pulse
- Quality of the trace
- Misinterpretation of CTG
- Inappropriate action with abnormal CTG
- Overall clinical picture and pattern evolution of CTG
- Role of infection and inflammation
- Team work and communication of findings
- CTG storage
- Training in CTG interpretation
- Documentation of events and risk management.

1. Patient Identity

The name of the woman, the date and time of commencement of recording should be entered on every trace. The clock on the machine must always be checked.

Table 23.1: Systematic interpretation of CTG trace[16]
Determine risk
Contraction frequency and duration
Baseline rate
Accelerations
Variability
Decelerations
Overall impression and care plan

Table 23.2: Description of cardiotocograph trace features (based on NICE guidance)[15]

Principles for intrapartum CTG trace interpretation

- When reviewing the CTG trace, assess and document contractions and all 4 features of FHR: baseline rate; baseline variability; presence or absence of decelerations (and concerning characteristics of variable decelerations if present); presence of accelerations.
- If there is a stable baseline FHR between 110 and 160 beats/minute and normal variability, continue usual care as the risk of fetal acidosis is low.
- If it is difficult to categorise or interpret a CTG trace, obtain a review by a senior midwife or a senior obstetrician.

Accelerations

The presence of FHR accelerations, even with reduced baseline variability, is generally a sign that the baby is healthy.

Description	Feature		
	Baseline (beats/minute)	**Baseline variability (beats/minute)**	**Decelerations**
Reassuring	110 to 160	5 to 25	None or early Variable decelerations with no concerning characteristics* for less than 90 minutes
Non-reassuring	100 to 109[†] or 161 to 180	Less than 5 for 30 to 50 minutes or More than 25 for 15 to 25 minutes	Variable decelerations with no concerning characteristics for 90 minutes or more or Variable decelerations with any concerning characteristics in up to 50% of contractions for 30 minutes or more or Variable decelerations with any concerning characteristics in over 50% of contractions for less than 30 minutes or Late decelerations in over 50% of contractions for less than 30 minutes, with no maternal or fetal clinical risk factors such as vaginal bleeding or significant meconium
Abnormal	Below 100 or Above 180	Less than 5 for more than 50 minutes or More than 25 for more than 25 minutes or Sinusoidal	Variable decelerations with any concerning characteristics in over 50% of contractions for 30 minutes (or less if any maternal or fetal clinical risk factors [See above]) or Late decelerations for 30 minutes (or less if any maternal or fetal clinical risk factors) or Acute bradycardia, or a single prolonged deceleration lasting 3 minutes or more

*Characteristics of variable decelerations include—decelerations lasting more than 60 seconds; reduced baseline variability within the deceleration; failure to return to baseline; biphasic (W) shape and no shouldering.

[†]Although a baseline fetal heart rate between 100 and 109 beats/minute is a non-reassuring feature, continue usual care if there is normal baseline variability and no variable or late decelerations.

Category	Definition	Management
Table 23.3: Management based on interpretation of cardiotocograph traces (NICE)[15]		
Normal	All features are reassuring	• Continue CTG (unless it was started because of concerns arising from intermittent auscultation and there are no ongoing risk factors and usual care • Talk to the woman and her birth companion(s) about what is happening
Suspicious	1 non-reassuring feature and 2 reassuring features	• Correct any underlying causes, such as hypotension or uterine hyperstimulation • Perform a full set of maternal observations • Start 1 or more conservative measures • Inform an obstetrician or a senior midwife • Document a plan for reviewing the whole clinical picture and the CTG findings • Talk to the woman and her birth companion(s) about what is happening and take her preferences into account

Pathological	1 abnormal feature
	or
	2 non-reassuring features
		• Obtain a review by an obstetrician and a senior midwife
		• Exclude acute events (for example, cord prolapse, suspected placental abruption or suspected uterine rupture)
		• Correct any underlying causes, such as hypotension or uterine hyperstimulation
		• Start 1 or more conservative measures
		• Talk to the woman and her birth companion(s) about what is happening and take her preferences into account
		• If the cardiotocograph trace is still pathological after implementing conservative measures:
		  • obtain a further review by an obstetrician and a senior midwife
		  • offer digital fetal scalp stimulation and document the outcome
		• If the cardiotocograph trace is still pathological after fetal scalp stimulation:
		  • consider fetal blood sampling
		  • consider expediting the birth
		  • take the woman's preferences into account

Need for urgent intervention	Acute bradycardia, or a single prolonged deceleration for 3 minutes or more	• Urgently seek obstetric help • If there has been an acute event (for example, cord prolapse, suspected placental abruption or suspected uterine rupture), expedite the birth • Correct any underlying causes, such as hypotension or uterine hyperstimulation • Start 1 or more conservative measures • Make preparations for an urgent birth • Talk to the woman and her birth companion(s) about what is happening and take her preferences into account • Expedite the birth if the acute bradycardia persists for 9 minutes • If the fetal heart rate recovers at any time up to 9 minutes, reassess any decision to expedite the birth, in discussion with the woman

If there are any concerns about the baby's well-being, be aware of the possible underlying causes and start one or more of the following conservative measures based on an assessment of the most likely cause(s): encourage the woman to mobilise or adopt an alternative position (and to avoid being supine); offer intravenous fluids if the woman is hypotensive; reduce contraction frequency by reducing or stopping oxytocin if it is being used and/or offering a tocolytic drug (a suggested regimen is subcutaneous terbutaline 0.25 mg).

Table 23.4: The classification of fetal blood sampling (FBS) results (NICE)[15]

Lactate (mmol/l)	pH	Interpretation
≤ 4.1	≥ 7.25	Normal
4.2–4.8	7.21–7.24	Borderline
≥ 4.9	≤ 7.20	Abnormal

These results should be interpreted considering the previous pH measurement, the rate of progress in labour and the clinical features of the woman and baby.

2. Maternal Pulse

FHR should be auscultated prior to application of the electronic probe to avoid picking up maternal pulsations. In addition, the maternal pulse should be identified and recorded separately. Any sudden significant shift in the baseline FHR or a wide variability would suggest recording of the maternal pulse rather than FHR. If there is doubt, ultrasound should be used to locate the fetal heart and a fetal scalp electrode may be a better alternative in such a situation. Slippage of transducer from tracking the fetal heart to maternal pulse during labour is not common but can happen. The whole CTG must be reviewed from time to time for sudden changes in the rate and special attention has to be paid during twin delivery after the first baby is born. In the second stage of labour, accelerations that coincide with contractions are likely to be maternal heart rate recordings.

3. Quality of the Trace

The FHR tracing is difficult to interpret when there is persistent signal loss. The situation should be corrected by adjusting the transducer, or obtaining the signal via a scalp electrode, or changing the connections and/or machine. If these actions do not rectify the problem, intermittent auscultation should be performed and this should be documented in the medical records.

4. Misinterpretation of CTGs

While interpreting a CTG trace, emphasis should be paid to observe for reactivity (accelerations) and cycling (quiet and active sleep cycles) that indicates a non-hypoxic fetus with a normal behavioural pattern. Absence of cycling may be due to drugs, infection, cerebral hemorrhage, chromosomal or congenital malformation, previous brain damage. A nonreactive trace with baseline variability <5 bpm and shallow decelerations (<15 beats) that lasts for >90 minutes suggests an existing hypoxia. In the presence of a clinical picture like post-term, growth restriction, absent fetal movements, antepartum hemorrhage or infection-such a trace should prompt earlier delivery. A previously brain damaged fetus may or may not show cycling but the cord pH at birth may be normal; such babies may not show evidence of HIE but may exhibit signs of neurological damage that manifests later. Presence of accelerations, normal baseline heart rate, variability more than 5 bpm and absence of any decelerations are features of a normal reassuring CTG (Fig. 23.1). With a normal baseline CTG, a gradually developing hypoxia will be reflected by no accelerations, repeated decelerations and gradually rising baseline rate (Fig. 23.2). Furthermore, it is

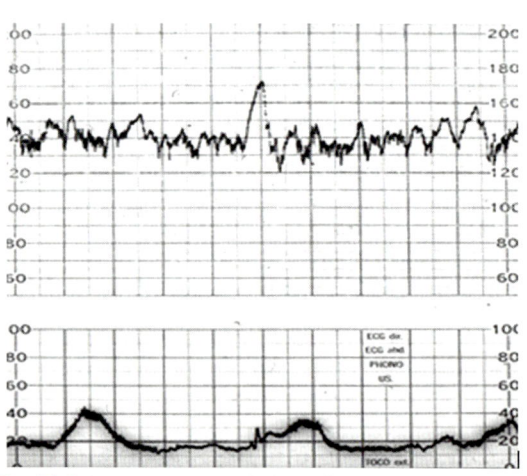

Fig. 23.1: Normal reactive CTG with accelerations, normal baseline, good variability without any decelerations.

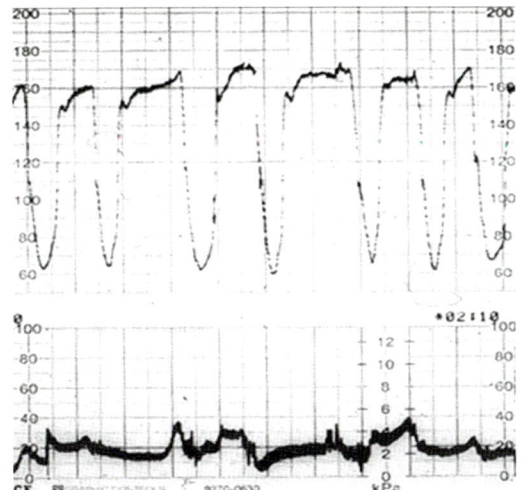

Fig. 23.2: CTG with baseline rate of 170 bpm, absent variability and atypical variable decelerations.

known that if a well grown fetus with clear amniotic fluid and a reactive CTG trace starts to develop an abnormal FHR pattern, it takes some time with these FHR changes before acidosis develops. A study estimated that in situations with abnormal FHR pattern—for 50% of the babies to become acidotic took 115 minutes with repeated late decelerations, 145 minutes with repeated variable decelerations and 185 minutes with a flat trace.[17] Fetuses with a reactive admission CTG will show following features prior to or becoming hypoxic—all will exhibit decelerations (100%), almost all will have reduced baseline variability (93%) and baseline tachycardia (93%).[18] A rising baseline associated with reduced variability therefore may be an ominous sign of fetal hypoxia where the fetus tries to increase oxygen delivery to vital organs by increasing cardiac output. On the other hand, if the baseline CTG is non-reactive, the development of further abnormal features with progress of labour are variable and subtle; this is difficult to recognise by intermittent auscultation. This is because there might be pre-existing hypoxic damage and the fetus is unable to respond. Such a fetus may not withstand the stress of uterine contractions and runs the risk of death within a few hours of admission.

When the baseline FHR is above 180 beats/min with no other non-reassuring or abnormal features on the cardiotocograph, possible underlying causes (such as infection) should be sought for. Woman's temperature and pulse should be checked and if either was raised, fluids and paracetamol should be offered. Fetal blood sampling to measure lactate or pH should be considered if the rate stays above 180 beats/min despite conservative measures.

Baseline variability will usually be 5 beats/minute or more. Intermittent periods of reduced baseline variability are normal, especially during periods of quiescence ('sleep'). Mild or minor pseudosinusoidal patterns (oscillations of amplitude 5–15 beats/minute) are also of no significance.

5. Inappropriate Action with Abnormal CTG

Once a diagnosis of suspicious or pathological FHR trace is made—action must be taken depending on the severity of CTG abnormality. Accurate documentation of the time of observation and any other actions taken is very important from a medicolegal view point.

In general, a 'suspicious CTG' can be managed conservatively. An 'pathological CTG' requires institution of conservative measures and fetal scalp stimulation or fetal scalp blood sampling (FBS) where appropriate/feasible, otherwise delivery should be expedited. Interventions for suspicious or pathological trace will depend on the suspected underlying cause for the abnormalities in FHR. These features may be reversed by conservative measures, such as changing maternal position, treating hypotension or pyrexia, hydration, reducing or stopping oxytocin or considering tocolysis for hyperstimulation.

Timely action is crucial in cases of prolonged decelerations or bradycardia for longer than 3 minutes where urgent intervention may be required. Probable causes include placental abruption, cord prolapse and uterine scar rupture. In the event of such emergency, immediate delivery should occur. In the absence of these, the '3, 6, 9, 12, 15 minutes' guidance can be followed. Interventions such as cessation of oxytocin and treatment of maternal hypotension should commence. If there are no signs of recovery at 6 minutes, preparations should be made to transfer to theatre by 9 minutes. caesarean section should commence by 12 minutes with the aim of delivery by 15 minutes. If instrumental delivery is possible this should be achieved within 15–20 minutes but a difficult instrumental delivery should be avoided. Experienced neonatal team should attend the delivery if resuscitation is anticipated. Many cases will recover by 9 minutes (especially if they had a normal CTG prior to this event) and cesarean may not be necessary unless there are additional reasons for concern. If the FHR recovers at any time up to 9 minutes, decision to expedite the birth should be reassessed, in discussion with the woman. Appropriate debriefing and explanation of events to the mother and partner should follow expedited delivery. Some of the CTG patterns pose a challenge to the obstetrician regarding their interpretation and management. The following features in isolation are unlikely to be associated with significant acidosis:[18] (a) Baseline fetal heart fate between 100 and 109, (b) Uncomplicated tachycardia (161–180) with accelerations present, (c) Absence of accelerations and (d) Variable decelerations without complicating features.

If the CTG trace is pathological, digital fetal scalp stimulation should be offered. If this leads to an acceleration in FHR, fetal blood sampling should only be considered if the trace remains pathological. FBS is recommended with pathological CTGs to identify those fetuses that require delivery and ideally all maternity units where CTG is employed should have ready access, 24 hours a day to an accurate blood gas analyser. Contraindications to FBS or invasive fetal monitoring are: maternal infections like HIV, Hepatitis B and C, active Herpes Simplex, fetal bleeding disorders, hemophilia (male fetus in carrier), preterm gestation < 34 weeks. Table 23.4 summarises the normal and abnormal acid–base values. After an abnormal FBS result, consultant obstetric advice should be sought. After a normal FBS result, sampling should be repeated no more than 1 hour later if the FHR trace remains pathological, or sooner if there are further abnormalities. After a borderline FBS result, sampling should be repeated no more than 30 minutes later if the FHR trace remains pathological or sooner if there are further abnormalities. The time taken to take a fetal blood sample needs to be considered when planning repeat samples. If the FHR trace remains unchanged and the FBS result is stable after the second test, a third/ further sample may be deferred unless additional abnormalities develop on the trace. Where a third FBS is considered necessary or sampling fails, consultant obstetric opinion should be sought as based on the clinical situation delivery may be more appropriate. The clinical context of the labour should always be considered including parity, progress, stage of labour and maternal wishes. It should be remembered that fetal infection or thick meconium associated with a pathological CTG may result in an adverse neonatal outcome even in the absence of fetal acidosis. A normal FBS result in these circumstances does not give the same reassurance. If a fetal blood sample is indicated and the sample cannot be obtained, but the associated scalp stimulation results in FHR accelerations, the clinician should decide whether to continue the labour or expedite the birth in light of the clinical circumstances and in discussion with the woman. If a fetal blood sample is indicated but a sample cannot be obtained and there is no improvement in the cardiotocograph trace, the woman should be advised that the birth should be expedited. Fetal lactate measurement has been proposed as an alternative to scalp pH as the lactate levels reflect anaerobic respiration and thus tissue hypoxia and metabolic acidosis. The invasive procedure for sampling is similar to FBS but requires a smaller volume of blood (5 microl) hence failure rates to obtain a sample is low (*See* Table 23.4).

6. Overall Clinical Picture and Pattern Evolution of CTG

Generally, during the intrapartum period a hypoxia-induced reduction in FHR variability develops gradually (commonly over approximately 60 minutes) and occurs in the context of recurrent late, variable, or prolonged decelerations. It is important therefore to be able to appreciate the CTG pattern evolution and recognise the gradual changes in FHR pattern tracing over time. Comparison of a trace at given point in time with another part of the trace some time back in the same fetus can give vital information, which can assist in decision-making.[19]

7. Role of Infection and Inflammation

Recently, the presence of infection/pyrexia has been found to be an important finding in fetuses that are destined to develop cerebral palsy.[19] The fetal inflammatory response associated with maternal fever during labour, choriamnionitis and funisitis has been implicated as a cause of later cerebral palsy. It is believed that inflammatory cytokines can cause cerebral ischemia resulting in damage to the fetal central nervous system. This information needs to be factored in while managing cases with pathological CTGs in relevant clinical scenarios.

8. Team Work and Communication of Findings

Effective intrapartum FHR monitoring requires good teamwork. All members of the maternity team (doctors, midwives, nurses) should be aware of how FHR traces are interpreted, which FHR patterns are associated with actual or impending fetal acidemia and within what time frame the senior team member should be notified of abnormal FHR pattern.

9. Storage of CTG

CTGs should be stored for at least 25 years and the hospital should make adequate provision for safe storage and easy retrieval. The CTGs are recorded on thermosensitive paper and tends to fade in 3 to 4 years' time. Where possible, traces should be stored electronically.

10. Training in CTG Interpretation

It is essential that all maternity units provide a regular and structured programme on interpretation of CTGs for all midwives and doctors working on the labour ward. Participation in weekly case review meetings and discussions on CTG traces is one of the best ways of reinforcing knowledge.

11. Documentation of Events and Risk Management

The role of clinical risk management is not to apportion blame but to improve the standard of clinical care and avoid adverse outcomes. The process involves identification/reporting of critical adverse incidents, root cause analysis of events as well as individual actions and finally recommendations to prevent or better manage similar incidents in future. Lessons learned from analysis of adverse obstetric events should be disseminated to all staff working on the labour ward.

CONCLUSION

Obstetric litigation is on the rise and a review of litigation cases reveals that majority of claims relating to the intrapartum period arise because of misinterpretation of CTGs or inappropriate action following abnormal CTG. Mandatory education and training in the interpretation of CTGs and best practice recommendations for intrapartum care are key factors to avoid litigations, and improve outcomes for mothers and babies.

REFERENCES

1. Williams B, Arulkumaran S. Cardiotocography and medico-legal issues. Best Pract Res Clin Obstet Gynaecol 2004;18(3):457–66.
2. Symonds EM, Senior EO. The anatomy of obstetric litigation. Current Opinion in Obstetrics and Gynecology 1991;1:241–3.
3. Freeman RK. Problems with intrapartum fetal heart rate monitoring interpretation and patient management. Obstet Gynecol 2002;100(4):813–26.
4. Nelson KB, Dambrosia JM, Ting TY, Grether JK. Uncertain value of fetal heart rate monitoring in predicting cerebral palsy. N Engl J Med 1996;334:613–8.
5. Hepworth S. Clinical cases by speciality. NHS Litigation Authority Journal 2003; Summer (Issue 2):4. [4] Ten Years of Maternity Claims. An Analysis of NHS Litigation Authority Data. NHS Litigation Authority. October 2012.
6. Fox M, Kilpatrick S, King T, Parer JT. Fetal heart rate monitoring: interpretation and collaborative management. J Midwifery Womens Health 2000;45(6):498–507.
7. Antenatal diagnosis. Report of a consensus development conference. NIH publication no. 79–1973. Bethesda, MD: National Institutes of Health, 1979.
8. American College of Obstetricians and Gynecologists. Fetal and neonatal neurologic injury. ACOG technical bulletin no. 163. Washington DC: American College of Obstetricians and Gynecologists, 1992.
9. Policy statement of the Task Force on Cerebral Palsy and Neonatal Asphyxia of the Society of Obstetricians and Gynecologists of Canada (part I). J Soc Obstet Gynecol Can 1996;1267–79.
10. MacLennan A. A template for defining a causal relation between acute intrapartum events and cerebral palsy: international consensus statement. BMJ 1999;319:1054–9.
11. American College of Obstetricians and Gynecologists and American Academy of Pediatrics. Neonatal encephalopathy and cerebral palsy: defining the pathogenesis and pathophysiology. Washington DC: American College of Obstetricians and Gynecologists; 2003.
12. Fatemi A, Wilson MA, Johnston MV. Hypoxic Ischemic Encephalopathy in the Term Infant. Clin Perinatol 2009;36(4):835–vii.doi:10.1016/j.clp.2009.07.011.
13. Heinz ER, Provenzale JM. Imaging findings in neonatal hypoxia: a practical review. AJR Am J Roentgenol 2009;192(1):41–7.doi:10.2214/AJR.08.1321. Sonography still play a role? Pediatr Radiol 2006;36:636–46.
14. Myers RE. Four patterns of perinatal brain damage and their conditions of occurrence in primates. Adv Neurol 1975;10:223–34.
15. NICE clinical guideline CG190. Intrapartum care for healthy women and babies (CG190). National Institute for Health and Care Excellence. 2014, updated February 2017, www.nice.org.uk.
16. Bailey RE, Hinshaw K. Intrapartum fetal monitoring. In: ALSO provider manual. 3rd edn. Kansas: AAFP, 2001.
17. Fleischer A, Schulman H, Jagani N, Mitchell J, Randolph G. The development of fetal acidosis in the presence of an abnormal fetal heart rate tracing. I. The average for gestational age fetus. Am J Obstet Gynecol 1982;144(1):55–60.
18. Leslie K, Arulkumaran S. Intrapartum fetal surveillance. Obstetrics, Gynaecology & Reproductive Medicine, 2011;Vol 21,Issue 3:59–6.
19. Sinai Talaulikar V, Arulkumaran S. Persistent Challenge of Intrapartum Fetal Heart Rate.

CHAPTER

24

NICE Guidelines for CTG Interpretation

Bini Ajay, Michelle Mooy, Sonali Gaur

One of the most common factors that contribute to fetal intrapartum death is related to the Cardiotocograph (CTG). This can involve a combination of an inability to interpret the CTG trace; failure to incorporate the clinical picture; delay in taking action and poor teamwork.

The incidence of Hypoxic Ischemic Encephalopathy (HIE) owing to birth asphyxia occurs in approximately 2 per 1000 live full-term births. Of affected newborns, 15–20% will die in the postnatal period and an additional 25% will develop severe and permanent neuropsychological sequelae, including mental retardation, visual motor or visual perceptive dysfunction, hyperactivity, cerebral palsy and epilepsy. The outcomes of HIE are devastating and permanent, making it a major burden for the patient, the family, and society.[1] Magnetic resonance imaging studies of babies born at term who develop cerebral palsy have revealed that up to 28% of cases may be attributable to intrapartum related asphyxia injuries.[2]

INDICATIONS FOR THE USE OF CONTINUOUS MONITORING IN LABOUR[3]

Do not offer cardiotocography to women at low-risk of complications in established labour. Advise continuous cardiotocography if any of the following risk factors are present at the initial assessment or arise during labour:

- Maternal pulse over 120 beats/minute on 2 occasions 30 minutes apart

- Temperature of 38°C or above on a single reading, or 37.5°C or above on 2 consecutive occasions 1 hour apart
- Suspected chorioamnionitis or sepsis
- Pain reported by the woman that differs from the pain normally associated with contractions
- The presence of significant meconium (as defined in ongoing assessment)
- Fresh vaginal bleeding that develops in labour
- Severe hypertension: A single reading of either systolic blood pressure of 160 mm Hg or more or diastolic blood pressure of 110 mm Hg or more, measured between contractions
- Hypertension: Either systolic blood pressure of 140 mm Hg or more or diastolic blood pressure of 90 mm Hg or more on 2 consecutive readings taken 30 minutes apart, measured between contractions
- A reading of 2+ of protein on urinalysis and a single reading of either raised systolic blood pressure (140 mm Hg or more) or raised diastolic blood pressure (90 mm Hg or more)
- Confirmed delay in the first or second stage of labour (*see* suspected delay in established first stage and delay in second stage)
- Contractions that last longer than 60 seconds (hypertonus), or more than 5 contractions in 10 minutes (tachysystole)
- Oxytocin use.

To ensure accurate record keeping for cardiotocography:
- Make sure that date and time clocks on the cardiotocograph monitor are set correctly.
- Make sure CTG paper speed is at 1 cm/minute.
- Label traces with the woman's name, date of birth and hospital number, the date and the woman's pulse at the start of monitoring.
- Individual units should develop a system for recording relevant intrapartum events (for example, vaginal examination, fetal blood sampling and siting of an epidural) in standard notes and/or on the cardiotocograph trace.
- Keep cardiotocograph traces for 25 years and, if possible, store them electronically.
- In cases where there is concern that the baby may experience developmental delay, photocopy cardiotocograph traces and store them indefinitely in case of possible adverse outcomes.
- Ensure that tracer systems are available for all cardiotocograph traces if stored separately from the woman's records (Fig. 24.1).

Interpreting the CTG

A useful pneumonic to use is **DR C BRAVADO.** The CTG should be analysed in each category, namely:

DR: Define risks, **C**: Contractions, **BRa**: Baseline rate, **V**: Variability, **A**: Accelerations, **D**: Decelerations, **O**: Overall impression.

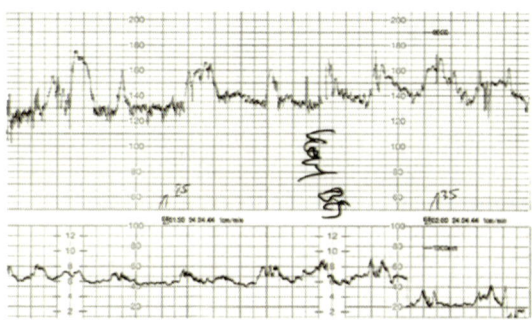

Fig. 24.1: Normal CTG with all reassuring features.

Define Risks

Always interpret the CTG in the context of the patient's clinical picture and begin by defining maternal and fetal risk factors of the particular case, e.g. prolonged rupture of membranes, intrauterine growth restriction, preterm or postdates pregnancy, patient on oxytocin infusion.

Intrauterine infection leads to a rise in perinatal morbidity and mortality and, in this context, the CTG is less sensitive at detecting hypoxia.[4] Always interpret the CTG with caution in the presence of maternal pyrexia which often accompanies a maternal tachycardia.

Meconium is strongly associated with histological chorioamnionitis, and in one study the RR of fetal infection was $\geq$ 50 fold if FHR tachycardia was associated with meconium in early labour.[5] Thick meconium with scanty fluid should be managed with caution.

Neonatal Meconium Aspiration Syndrome (MAS) is the result of aspiration of acidic meconium causing chemical pneumonitis. This causes cytokine aspiration and can lead to the development of persistent pulmonary hypertension. Any developing hypoxia increases the risk of fetal gasping *in utero* and this increases the risk of MAS.[6]

The CTG will be interpreted more cautiously in the presence of risk factors and the threshold for intervention will be lower than in a low-risk case.

Contractions

Ascertain the number (frequency) of contractions in a 10 minute period, e.g. normally 3 or 4 in 10. The duration of a contraction is the interval between the time when the uterus is first observed to contract, to when the pressure has returned to the resting level.

Baseline Rate

Next, ascertain the baseline Fetal Heart Rate (FHR). This is the mean level of FHR excluding accelerations and decelerations:
- Reassuring: 110 to 160 beats/minute

- Non-reassuring: 100 to 109 beats/minute 161 to 180 beats/minute
- Abnormal: Below 100 beats/minute above 180 beats/minute.

Take the following into account when assessing baseline FHR:

- Differentiate between fetal and maternal heartbeats
- Baseline FHR will usually be between 110 and 160 beats/minute
- Although a baseline FHR between 100 and 109 beats/minute is a non-reassuring feature, continue usual care if there is normal baseline variability and no variable or late decelerations.

Accelerations

An acceleration is a visually apparent abrupt increase in FHR. The peak must be ≥ 15 bpm and the acceleration must last for at least 15 seconds before returning to baseline.[7]

The presence of accelerations, even with reduced baseline variability, is generally a sign that the fetus is healthy.[8] The absence of accelerations in an otherwise normal cardiotocograph trace does not indicate fetal acidosis (Fig. 24.2).

If digital fetal scalp stimulation (during vaginal examination) leads to an acceleration in FHR, regard this as a sign that the baby is healthy. Take this into account when reviewing the whole clinical picture.

Variability

Baseline FHR variability is defined as fluctuations in the baseline FHR that are irregular in amplitude and frequency. This is quantified as the amplitude of peak to trough in beats per minute.

Moderate baseline FHR variability reflects the delivery of oxygen to the fetal central nervous system. Its presence is reassuring in predicting an absence of metabolic acidemia and hypoxic injury to the fetus at the time it is observed. In contrast, the presence of mild baseline FHR variability, or an absence of FHR variability does not reliably predict fetal acidemia or hypoxia; lack of moderate baseline FHR variability may be a result of the fetal sleep cycle, or a result of medications, extreme prematurity, congenital anomalies, or pre-existing neurological injury.[9]

Use the following categorisations for FHR baseline variability:

- **Reassuring**: 5 to 25 beats/minute (Fig. 24.3)

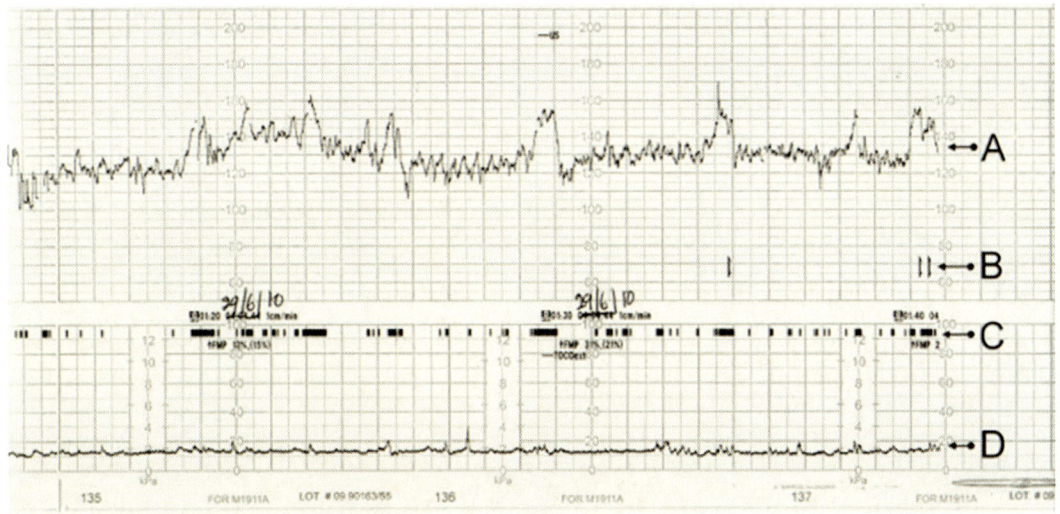

Fig. 24.2: CTG trace with accelerations.

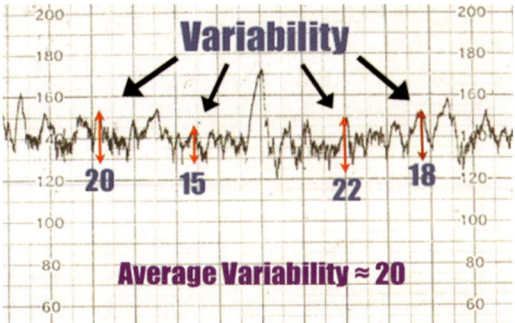

Fig. 24.3: Reassuring variability (5–25 bpm).

- **Non-reassuring**: Less than 5 beats/minute for 30 to 50 minutes or more than 25 beats/minute for 15 to 25 minutes
- **Abnormal**: Less than 5 beats/minute for more than 50 minutes or more than 25 beats/minute for more than 25 minutes; **sinusoidal.**

Take the following into account when assessing FHR baseline variability:

- Baseline variability will usually be between 5 and 25 beats/minute
- Intermittent periods of reduced baseline variability are normal, especially during periods of quiescence ('sleep').

- Cycling with a reactive trace followed by sleep pattern suggests that the baby is neurologically normal.

Sinusoidal Pattern

A sinusoidal FHR pattern is defined as a pattern of fixed, uniform fluctuations of the FHR that creates a pattern resembling successive geometric sine waves. It frequently is described as undulating and smooth and is characterized by the absence of variability.

- *Physiological:* Thumbsucking/rhythmic fetal mouth movements (Fig. 24.4)

- *Pathological:* Defined as stable baseline 110 to 150 bpm with regular oscillations of amplitude 5 to 15 bpm at frequency of 2 to 5 cycles per minute with fixed/flat baseline variability (Figs 24.5A and B). If lasting >30 minutes, it can be associated with:

  • Fetal anemia
  • Fetal hypoxia
  • Fetal infection.

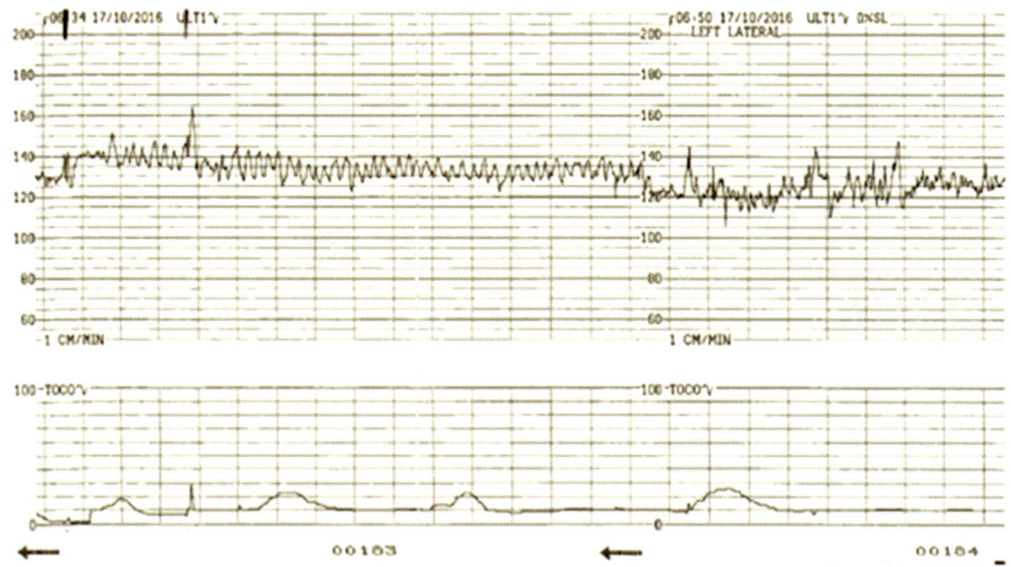

Fig. 24.4: Sinusoidal pattern (Physiological) (note change to normal variability).

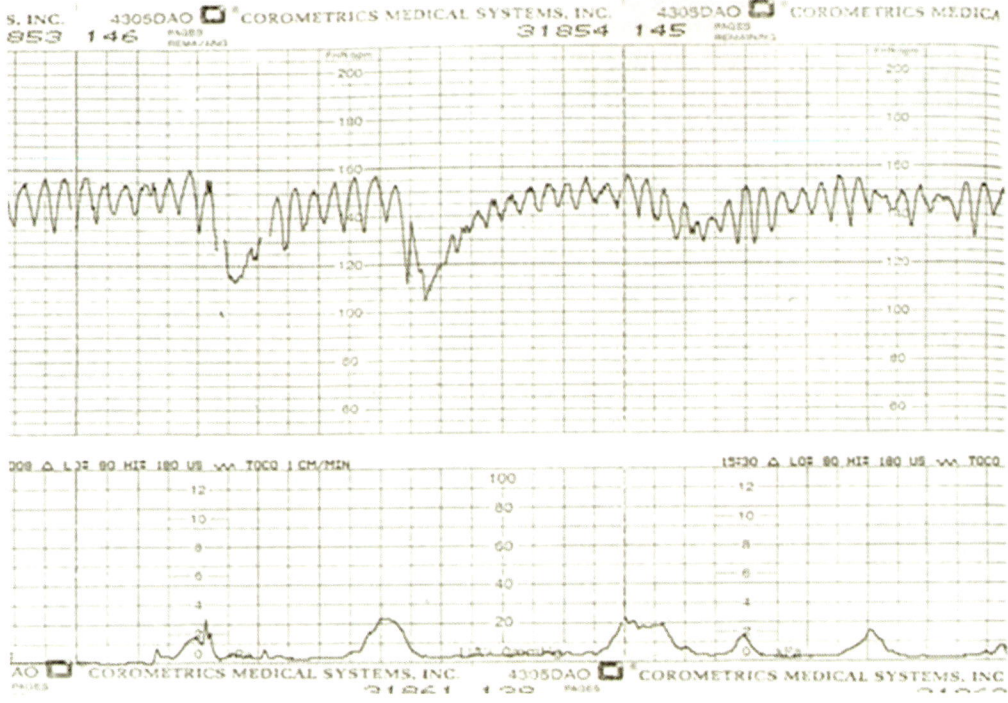

Fig. 24.5A: Sinusoidal with decelerations (Pathological).

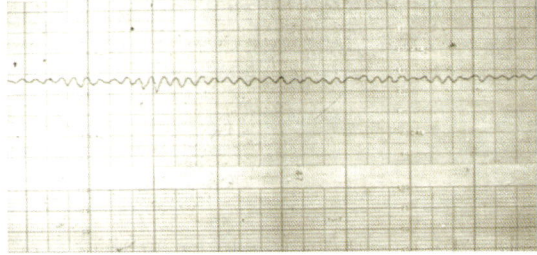

Fig. 24.5B: Sinusoidal pattern (sine wave pattern).

Decelerations

Describing Decelerations

When describing decelerations in FHR, specify:

- Their timing in relation to the peaks of the contractions; the duration of the individual decelerations, whether or not the FHR returns to baseline; how long they have been present for.
- Early decelerations—late variable

- Whether they occur with over 50% of contractions; the presence or absence of a biphasic (W) shape
- The presence or absence of shouldering
- The presence or absence of reduced variability within the deceleration.
- Describe decelerations as 'early', 'variable' or 'late'. Do not use the terms 'typical' and 'atypical' because they can cause confusion.

Categorising Decelerations

Use the following categorisations for decelerations in FHR:

- **Reassuring**: No decelerations; early decelerations; variable decelerations with no concerning characteristics (Fig. 24.6A) for less than 90 minutes
- **Non-reassuring**:
  - Variable decelerations with no concerning characteristics for 90 minutes or more

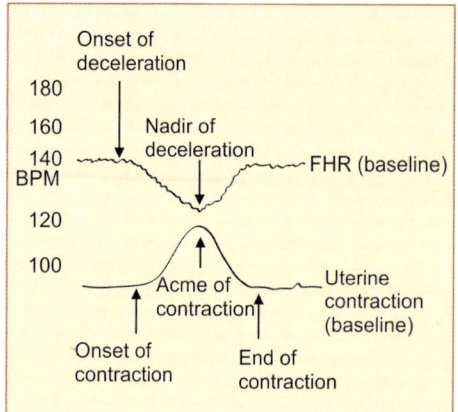

Fig. 24.6A: Early deceleration (nadir with contraction).

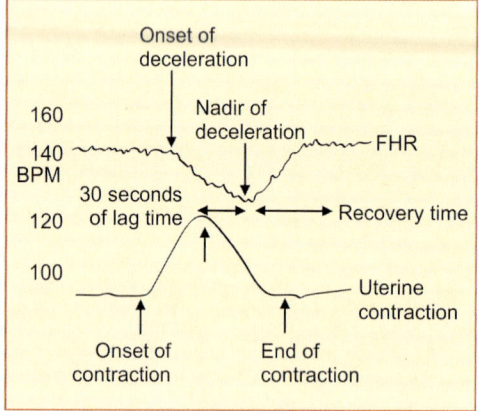

Fig. 24.6B: Late Deceleration (Nadir >20 seconds post contraction).

- Variable decelerations with any concerning characteristics in up to 50% of contractions for 30 minutes or more
- Variable decelerations with any concerning characteristics in over 50% of contractions for less than 30 minutes
- Late decelerations in over 50% of contractions for less than 30 minutes, with no maternal or fetal clinical risk factors such as vaginal bleeding or significant meconium (Fig. 24.6B).

■ **Abnormal:**
- Variable decelerations with any concerning characteristics in over 50% of contractions for 30 minutes (or less if there are any maternal or fetal clinical risk factors)

- Late decelerations for 30 minutes (or less if there are any maternal or fetal clinical risk factors)
- Acute bradycardia, or a single prolonged deceleration lasting 3 minutes or more.

Variable Decelerations

Regard the following as concerning characteristics of variable decelerations:

■ Lasting more than 60 seconds; reduced baseline variability within the deceleration; failure to return to baseline; biphasic (W) shape; no shouldering
■ There are different types of variable decelerations (Figs 24.7 to 24.10)
■ If variable decelerations with no concerning characteristics (*See* previous recommendation) are observed:
- Be aware that these are very common and can be a normal feature in an otherwise uncomplicated labour and birth. They are usually a result of cord compression. Ask the woman to change position or mobilise.

Assessing Decelerations

Take the following into account when assessing decelerations in FHR:
■ Early decelerations are uncommon, benign and usually associated with head compression
■ Early decelerations can be worrying if woman is not in labour or in very early labour with head not in pelvis

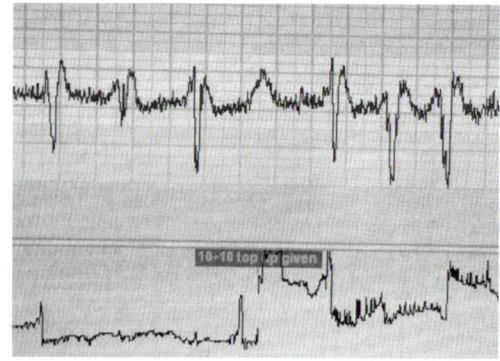

Fig. 24.7: Typical variable deceleration.

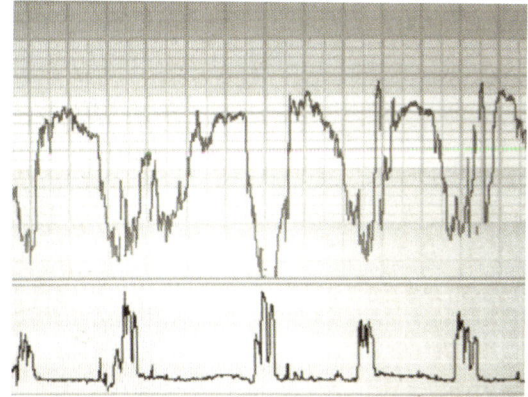

Fig. 24.8: Atypical (Biphasic) variable deceleration.

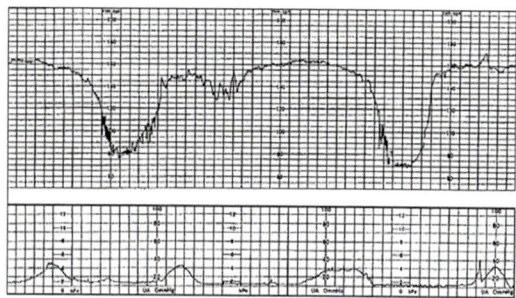

Fig. 24.9: Complicated Atypical variable deceleration.

- Early decelerations with no non-reassuring or abnormal features on the cardiotocograph trace should not prompt further action
- Take into account that the longer and later the individual decelerations, the higher the risk of fetal acidosis (particularly if the decelerations are accompanied by tachycardia or reduced baseline variability)

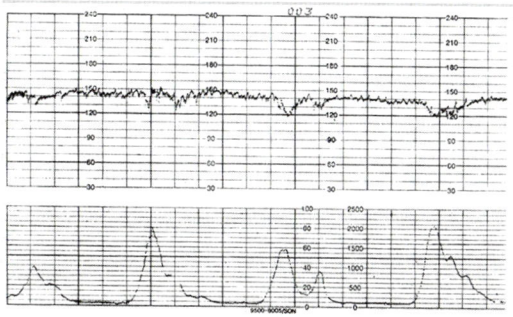

Fig. 24.10: Reduced variability with shallow deceleration – Ominous trace.

- Note that shallow (<15 bpm) late decelerations in the presence of reduced variability is an ominous sign.

Overall Assessment and Plan of Management

Categorise cardiotocography traces as follows:

- **Normal:** All features are reassuring *see* description of cardiotocograph trace features
- **Suspicious:** One non-reassuring feature and two reassuring features (but note that if accelerations are present, fetal acidosis is unlikely)
- **Pathological:** One abnormal feature or two non-reassuring features.

If continuous cardiotocography has been started because of concerns arising from intermittent auscultation, but the trace is normal after 20 minutes, return to intermittent auscultation unless the woman asks to stay on continuous cardiotocography.

If there is a stable baseline FHR between 110 and 160 beats/minute and normal variability, continue usual care as the risk of fetal acidosis is low.

If the cardiotocograph trace is categorised as normal: continue cardiotocography (unless it was started because of concerns arising from intermittent auscultation and there are no ongoing risk factors; *See* above) and usual care Talk to the woman and her birth companion(s) about what is happening.

Acute Bradycardia or Single Prolonged Deceleration

If there is an acute bradycardia, or a single prolonged deceleration for 3 minutes or more, urgently seek obstetric help (Fig. 24.11). If there has been an acute event (for example, cord prolapse, suspected placental abruption or suspected uterine rupture), expedite the birth (*See* expediting birth). Correct any underlying causes, such as hypotension or

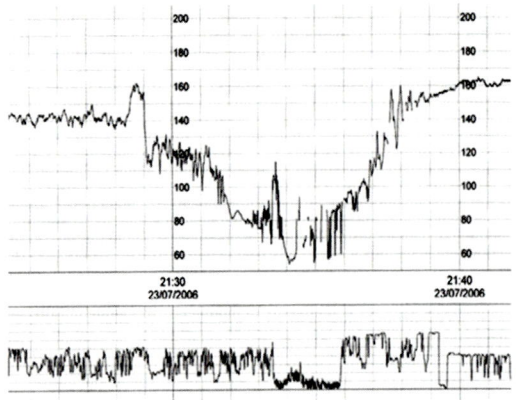

Fig. 24.11: Fetal Bradycardia.

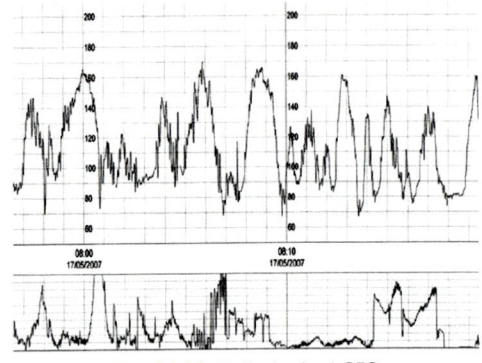

Fig. 24.12: Pathological CTG.

uterine hyperstimulation. Start one or more conservative measures and make preparations for an urgent birth. Remember to talk to the woman and her birth companion(s) about what is happening and take her preferences into account. Expedite the birth if the acute bradycardia persists for 9 minutes. If the FHR recovers at any time up to 9 minutes, reassess any decision to expedite the birth, in discussion with the woman.

Pathological Cardiotocograph Trace

If the cardiotocograph trace is categorised as pathological, obtain a review by an obstetrician and a senior midwife.

Exclude acute events (for example, cord prolapse, suspected placental abruption or suspected uterine rupture). Correct any underlying causes, such as hypotension or uterine hyperstimulation and start one or more conservative measures. Remember to talk to the woman and her birth companion(s) about what is happening and take her preferences into account. If the cardiotocograph trace is still pathological after implementing conservative measures, obtain a further review by an obstetrician and a senior midwife. Offer digital fetal scalp stimulation and document the outcome. If this leads to an acceleration in FHR, only continue with fetal blood sampling if the cardiotocograph trace is still pathological (Fig. 24.12).

If the cardiotocograph trace is still pathological after fetal scalp stimulation, consider fetal blood sampling (*See* fetal blood sampling) or expediting the birth (*See* expediting birth). Take the woman's preferences into account.

Suspicious Cardiotocograph Trace

If the cardiotocograph trace is categorised as suspicious, correct any underlying causes, such as hypotension or uterine hyperstimulation. Perform a full set of maternal observations and start one or more conservative measures. Inform an obstetrician or a senior midwife and document a plan for reviewing the whole clinical picture and the cardiotocography findings. Talk to the woman and her birth companion(s) about what is happening and take her preferences into account.

If there are any concerns about the baby's well-being, be aware of the possible underlying causes and start one or more of the following conservative measures based on an assessment of the most likely cause(s):

Encourage the woman to mobilise or adopt an alternative position (and to avoid being supine) and offer intravenous fluids if the woman is hypotensive. Reduce contraction frequency by reducing or stopping oxytocin if it is being used and/or offering a tocolytic drug (a suggested regimen is subcutaneous terbutaline 250 micrograms). Inform a senior midwife or an obstetrician whenever conservative measures are implemented.

Intrauterine Fetal Resuscitation

Do not use maternal facial oxygen therapy for intrauterine fetal resuscitation, because it may harm the baby (but it can be used where it is administered for maternal indications such as hypoxia or as part of preoxygenation before a potential anesthetic).

Do not offer amnioinfusion for intrauterine fetal resuscitation. If the cardiotocograph trace is pathological (*See* categorisation of traces), offer digital fetal scalp stimulation. If this leads to an acceleration in FHR, only continue with fetal blood sampling if the cardiotocograph trace is still pathological. If digital fetal scalp stimulation (during vaginal examination) leads to an acceleration in FHR, regard this as a sign that the baby is healthy. Take this into account when reviewing the whole clinical picture.

It is important to document your assessment of the CTG as well as your management plan. If it is a suspicious CTG document when you will next review it, typically in 20 minutes once implementing conservative measures if appropriate.

Common Mistakes

Below is the CTG which shows baseline of 160 bpm with variable decelerations. Sometimes it is confused with excessive fetal movements or baseline of 100 per minute with accelerations. Especially in second stage there will be no accelerations or fetal movements hence delivery is required urgently.

SUMMARY (Flow Chart 24.1)

Intrapartum FHR interpretation—stepwise physiologic approach:

- Step 1—the normal and the abnormal initial CTG
- Step 2—recognition of the compensated and the decompensating fetus.

Step 1

- If normal continue

- If the initial baseline FHR in a term fetus is ≥ 160 bpm with decelerations and reduced variability, particularly in association with meconium in early labour, the clinician should consider fetoplacental infection, meconium aspiration, chronic hypoxia, antecedent brain injury, maternal systemic disease, drugs, or chromosomal abnormality
- Senior staff involvement should be sought early, with consideration given to delivery by cesarean section.

Step 2

- An intact fetus with a previously normal CTG will exhibit predictable patterns of FHR responses if exposed to hypoxic ischemic insults during labour, namely:
  - Slowly evolving hypoxia
  - Subacute hypoxia
  - Acute hypoxia.

Slowly Evolving Hypoxia

- First intermittent episodes of oxygen deprivation will lead to decelerations
- The second is a progressive increase in baseline FHR
- The third is reduced variability, which is a marker of decompensation.

Subacute Hypoxia

This pattern is characterised by:
- Complicated variable decelerations with amplitude ≥ 60 bpm, duration ≥ 90 seconds, and a recovery phase at the baseline lasting < 60 seconds. This very brief interdeceleration interval is likely to have two consequences:
  - It is insufficient for the fetus to get rid of its carbon dioxide burden accumulated during the decelerations, leading rapidly to respiratory and subsequently metabolic acidosis
  - The fetus is unable to raise its baseline FHR and therefore its cardiac output.

Flow Chart. 24.1: Stepwise physiologic approach.

```
                        ┌─────────────────────────────┐
                        │   Initial intrapartum CTG   │
                        └─────────────────────────────┘
```

Normal	Abnormal Baseline FHR ≥ 160 bpm with decelerations ± mecanium ± reduced FHR variability

The fetus;
■ is normoxic, no acidosis
■ has normal acid base status
■ is not asphyxiated
■ has normal CNS and CVS
■ can react and defend itself
■ will exhibit predictable and
  progressive FHR changes if
  exposed to hypoxia ischemia
■ has low probability of
  intrapartum asphyxia
	Consider
	■ feto-placental infection
	■ meconium aspiration syndrome
	■ chronic hypoxia
	■ antecedent brain injury
	■ maternal systemic disease
	■ drugs
	■ chromosomal abnormality
	■ FBS not recommended

	Consider expeditious delivery

Continue surveillance or transfer to intermittent auscultation

a Repeated FHR decelerations > 50% of contractions b Acute prolonged FHR deceleration > 3 minutes

Compensated fetus	Compensated but 'stressed' fetus	Decompensating fetus
a ■ baseline FHR ≤ 160 bpm ■ FHR variability ≥ 5 bpm ■ deceleration amplitude ≤ 60 bpm ■ interval ≥ 60 s ■ ± cycling activity	a ■ baseline FHR ≥ or ≤ 160 bpm ■ FHR variability ≥ 5 bpm ■ deceleration amplitude ≥ 60 bpm ■ interval-deceleration interval ≥ 60 s ■ ± cycling activity ■ second test of fetal well-being	a ■ baseline FHR ≥ or ≤ 160 bpm ■ FHR variability ≥ 3–5 bpm ■ deceleration amplitude ≥ or ≤ 60 bpm ■ inter-deceleration interval < 60 s ■ duration of deceleration > 60s ■ second test of fetal well-being if appropriate

b
FHR 80–100 bpm for >3 minutes but
≤9 minutes with normal variability
recovery to normal or pre
deceleration CTG pattern
– exclude abruption, cord prolapse,
  uterine rupture, bolus of oxytocin
– remove cause ± expodite delivery
  if decompensation or failure to
  recover ≥ 10 minutes
	b
	FHR 80–100 bpm for > 3 minutes but
	≤ 9 minutes with normal variability recovery
	to FHR tachycardia or baseline FHR
	higher than pre deceleration CTG
	pattern ± unsuccessful recovery attemps
	– exclude abruption, cord prolapse,
	  uterine rupture, bolus of oxytocin
	– remove cause ± expodite delivery if
	  decompensation or failure to recover
	  ≥ 10 minutes
		b
		FHR ≤ 80–100 bpm for > 3 minutes with;
		– reduced FHR variability < 5 bpm
		– no signs of recovery observed
		– previously pathological FHR pattern
		– abrupt and erratic 'saltatory' pattern
		– no obvious cause identified or no
		  response to remedial action
		– consider expeditious delivery

Provided the FHR variability is normal and the interdeceleration interval ≥ 60 seconds the fetus is compensated

- However, subacute hypoxia is associated with a rapid decline in pH of 0.01 every 2–4 minutes
- Early recognition and remedial action is essential, as there may be insufficient time for further assessment, e.g. to obtain, analyse, and react to a fetal scalp sample result.

Acute Hypoxia (Prolonged FHR Deceleration)

- The majority of acute onset intrapartum FHR decelerations, in which the baseline FHR stabilises around 80–100 bpm, with normal variability, are associated with non-asphyxial vagal events.
- They usually arise from normal or near normal FHR patterns. In the absence of cord prolapse or occlusion, major abruption,

uterine rupture, maternal collapse, or infusion of a bolus of oxytocin, 90% of these episodes will recover or show signs of recovery by 6 minutes, and 95% will recover by 9 minutes

- If, however, the FHR falls < 80 bpm, with a loss of baseline variability, immediate delivery should be considered, especially if the antecedent CTG was abnormal as loss of variability signals fetal decompensation and injury.

Bradycardia of < 80 bpm can drop pH by 0.01 per minute.

REFERENCES

1. Ming-Chi Lai 1, San-Nan Yang. Perinatal Hypoxic-Ischemic Encephalopathy. J Biomed Biotechnol 2011;609813. Published online 2010. doi: 10.1155/2011/609813.

2. Hagberg B, Hagberg G, Beckung E, Uvebrant P. Changing panorama of cerebral palsy in Sweden. Prevalence and origin in the birth year period 1991–94. ActaPaediatr 2001;90:271–7.

3. NICE intrapartum care fetal monitoring page 5.

4. RCOG Strat OG—Electronic fetal monitoring.

5. Blot P, Milliez J, Breart G, Vige P, Nessmann C, Onufryk JP, et al. Fetal tachycardia and meconium staining: a sign of fetal infection. Int J Gynaecol Obstet 1983;21:189–94.

6. Fleischer A, Schulman H, Jagani N, Mitchell. Handbook of CTG interpretation From Patterns to Physiology Edited by Edwin Chandraharan. Chapter 13 Nirmala Chandrasekaran and Leonie Penna. Page 81.

7. Macones George A, Hankins Gary DV, Spong Catherine Y, Hauth John, Moore Thomas. "The 2008 National Institute of Child Health and Human Development Workshop Report on Electronic Fetal Monitoring". Obstetrics & Gynecology.

8. NICE intrapartum care Feb 2017.

9. Bailey RE. "Intrapartum fetal monitoring". American Family Physician 2009;80(12):1388–96.

Index